RETHINKING URBAN TRANSITIONS

Rethinking Urban Transitions provides critical insight for societal and policy debates about the potential and limits of low carbon urbanism. It draws on over a decade of international research, undertaken by scholars across multiple disciplines concerned with analysing and shaping urban sustainability transitions. It seeks to open up the possibility of a new generation of urban low carbon transition research, which foregrounds the importance of political, geographical and developmental context in shaping the possibilities for a low carbon urban future.

The book's contributions propose an interpretation of urban low carbon transitions as primarily social, political and developmental processes. Rather than being primarily technical efforts aimed at measuring and mitigating greenhouse gases, the low carbon transition requires a shift in the mode and politics of urban development. The book argues that moving towards this model requires rethinking what it means to design, practise and mobilize low carbon in the city, while also acknowledging the presence of multiple and contested developmental pathways. Key to this shift is thinking about transitions, not solely as technical, infrastructural or systemic shifts, but also as a way of thinking about collective futures, societal development and governing modes – a recognition of the political and contested nature of low carbon urbanism. The various contributions provide novel conceptual frameworks as well as empirically rich cases through which we can begin to interrogate the relevance of socio-economic, political and developmental dimensions in the making or unmaking of low carbon in the city. The book draws on a diverse range of examples (including 'world cities' and 'ordinary cities') from North America, South America, Europe, Australia, Africa, India and China, to provide evidence that expectations, aspirations and plans to undertake purposive socio-technical transitions are both emerging and encountering resistance in different urban contexts.

Rethinking Urban Transitions is an essential text for courses concerned with cities, climate change and environmental issues in sociology, politics, urban studies, planning, environmental studies, geography and the built environment.

Andrés Luque-Ayala is an Assistant Professor in the Department of Geography at Durham University, UK.

Simon Marvin is Director of the Urban Institute and Professor at the University of Sheffield, UK.

Harriet Bulkeley is Professor in the Department of Geography at Durham University, UK.

RETHINKING URBAN TRANSITIONS

Politics in the Low Carbon City

Edited by Andrés Luque-Ayala, Simon Marvin and Harriet Bulkeley

Routledge
Taylor & Francis Group

LONDON AND NEW YORK

First published 2018
by Routledge
2 Park Square, Milton Park, Abingdon, Oxon OX14 4RN

and by Routledge
711 Third Avenue, New York, NY 10017

Routledge is an imprint of the Taylor & Francis Group, an informa business

© 2018 selection and editorial matter, Andrés Luque-Ayala, Simon Marvin and Harriet Bulkeley; individual chapters, the contributors

The right of Andrés Luque-Ayala, Simon Marvin and Harriet Bulkeley to be identified as the authors of the editorial material, and of the authors for their individual chapters, has been asserted in accordance with sections 77 and 78 of the Copyright, Designs and Patents Act 1988.

British Library Cataloguing in Publication Data
A catalogue record for this book is available from the British Library

Library of Congress Cataloging in Publication Data
Names: Luque-Ayala, Andrés, editor. | Marvin, Simon, 1963- editor. | Bulkeley, Harriet, 1972- editor.
Title: Rethinking urban transitions : politics in the low carbon city / edited by Andrés Luque-Ayala, Simon Marvin and Harriet Bulkeley.
Description: Abingdon, Oxon ; New York, NY : Routledge, 2018. | Includes bibliographical references and index.
Identifiers: LCCN 2017052817 | ISBN 9781138057357 (hardback : alk. paper) | ISBN 9781138057401 (pbk. : alk. paper) | ISBN 9781315164779 (ebook)
Subjects: LCSH: City planning--Environmental aspects. | Urban ecology (Biology) | Climatic changes. | Carbon dioxide--Environmental aspects.
Classification: LCC HT166 .R4425 2018 | DDC 307.1/216--dc23
LC record available at https://lccn.loc.gov/2017052817

ISBN: 978-1-138-05735-7 (hbk)
ISBN: 978-1-138-05740-1 (pbk)
ISBN: 978-1-315-16477-9 (ebk)

Typeset in Bembo
by Taylor & Francis Books

This book is dedicated to Alex Aylett (1978–2016), a pioneer in the field whose dedication and enthusiasm for urban sustainability and justice is a continued inspiration to activists and researchers who want a better future.

CONTENTS

FIGURES

TABLES

CONTRIBUTORS

Editors

Andrés Luque-Ayala is an Assistant Professor in the Department of Geography at Durham University, UK. His research focuses on the politics of urban infrastructures in the global South. Currently he is working on the coupling of digital and material infrastructures as a new security apparatus in the city.

Simon Marvin is Director of the Urban Institute and Professor at the University of Sheffield, UK. He is an urban technologist who explores the reciprocal relationships between urban and socio-technical change. His current research explores the intersections between computational logic and its transmutation into urban contexts.

Harriet Bulkeley is Professor of Human Geography at Durham University, UK. Her research is concerned with environmental governance and focuses on theorizing and explaining the processes and practices of governing the environment, the urban politics of climate change and sustainability, and the political geographies of environmental governance.

Chapter contributors

Robyn Dowling is Associate Dean Research and Professor of Urbanism in the Sydney School of Architecture, Design and Planning at the University of Sydney, Australia. Her research explores the reshaping of urban governance in response to large-scale challenges like climate change and technological disruptions. Currently she is researching the governance impacts of smart mobility forms like car sharing and autonomous vehicles, and the ways in which smart city strategies are being implemented. She is an editor of *Transactions of the Institute of British Geographers.*

Maria Francesch-Huidobro is a political scientist consulting for the Regional Project Energy Security and Climate Change Asia–Pacific of the Konrad-Adenauer Stiftung. She is a certified member of the Environmental Management Association of Hong Kong and a professional member of the Hong Kong Institute of Qualified Environmental Professionals.

Mikael Granberg is Professor of Political Science and Director of the Centre for Climate & Safety (CCS), Karlstad University, Sweden. He is also a research fellow and board member of the National Strategic Research Area at the Centre for Natural Hazards and Disaster Science (CNDS), Sweden.

Mike Hodson is Senior Research Fellow in the Sustainable Consumption Institute, University of Manchester, UK. His research focuses on the shape of future sustainable cities and transitions in urban infrastructures. His current work, with colleagues from the Alliance Manchester Business School, analyses the implications of political devolution for urban transitions in housing and transport and how processes of devolution can be enhanced.

Ralph Horne is Professor of Geography and Deputy Pro-Vice-Chancellor, Research and Innovation, for the College of Design and Social Context at RMIT University, Australia. He is interested in social and policy change to support sustainable urban development. He is also Director of the Cities Programme, the urban arm of the United Nations Global Compact. He combines research leadership and participation in research projects concerning the environmental, social and policy context of production and consumption in the urban environment.

Pauline McGuirk joined the University of Wollongong, Australia, as Professor of Human Geography in 2016. As an urban political geographer, her current research investigates the material means and social practices through which energy transition governance is playing out through urban spaces, and the governance processes and practices involved in making cities smart in Australia. She is an editor of *Progress in Human Geography* and an elected fellow of the Academy of Social Sciences of Australia.

Andy McMeekin is Professor of Innovation at Alliance Manchester Business School, UK, and Research Director of the Sustainable Consumption Institute. Until recently he was co-director of the Economic and Social Research Council, Scottish Government, and DEFRA (Department for the Environment, Food and Rural Affairs)-funded Sustainable Practices Research Group (2011–14). His current research looks at long-term processes of socio-technical innovation, combing insights from economic sociology, practice theory and transition theory.

Susie Moloney is Senior Lecturer in Sustainability and Urban Planning and a researcher in the Centre for Urban Research, based in the School of Global, Urban and Social Studies at RMIT University, Australia. Her research focuses on land-use planning, urban sustainability and climate change, and the implications for policy and governance arrangements particularly at the local and regional scale. She has over ten years experience working in the public and private sectors on a range of planning and climate change-related projects.

Timothy Moss is Senior Researcher at the Integrative Research Institute for Transformations of Human–Environment Systems at the Humboldt University of Berlin, Germany. With a background in European Studies and History he has researched urban infrastructures as socio-technical configurations for twenty-five years, focusing on their vibrant histories, contested politics and multiple geographies. He is currently working on a book on the political history of Berlin's infrastructures from 1920 to the present.

Merissa Mueller completed her MSc in Environmental Sustainability at the University of Ottawa, Canada, in 2015. Her research expertise revolves around planning for intensification, focusing on the barriers to the redevelopment of inner-urban neighbourhoods and the role of public participation in these processes.

Matthew Paterson is Professor of International Politics at the University of Manchester, UK. His work focuses on the global governance, political economy and cultural politics of climate change. His works on this include *Climate Capitalism* (with Peter Newell, 2010) and *Transnational Climate Change Governance* (with Harriet Bulkeley and eight others, 2014).

Jonathan Rutherford holds a research post at LATTS (Laboratoire Techniques, Territoires et Sociétés), Université Paris Est, France. His research interests are in the processes and politics of urban socio-technical change through a focus on the shifting relations between infrastructure and cities. He co-edited recent special issues of *Urban Studies* and *Energy Policy*, and the Routledge volume, *Beyond the Networked City*.

Neha Sami has been faculty at the Indian Institute for Human Settlements in Bangalore since 2012, where she also anchors the institution's Research Programme. With a background in urban planning, environmental studies, and economics, her research focuses on the planning and governance arrangements of large infrastructure projects, and on environmental and climate governance questions in Indian cities. She also serves on the editorial collective of *Urbanisation*.

Jonathan Silver is a Leverhulme Early Career Fellow at the Urban Institute, University of Sheffield, UK. Jonathan is a geographer working on global urban studies and infrastructure through comparative research. He is particularly interested in new contested flows of carbon capital into cities, the tensions of urbanization and climate change and the resulting politics of the climate crisis. Much of his empirical focus has been on African cities including Accra, Cape Town and Kampala.

Laura Tozer is a Post-Doctoral Research Associate in the Department of Geography at Durham University, UK. Her research focuses on the policies and politics of climate change governance, urban sustainability, and low carbon transitions for built environments, infrastructure and energy.

ACKNOWLEDGEMENTS

First, we would like to acknowledge the support of the UK Economic and Social Research Council (ESRC) for funding the Urban Low Carbon Transitions Network: A Comparative International Network (Grant Reference ES/J019607/1). This made possible the development of a four-year programme of sustained interaction between the network members. All of the network members and additional authors are represented in the book project. Although the funding was only designed to support UK-based network meetings, we were able, due to the generosity and support of network members, to host international meetings in the UK, Hong Kong, Taiwan, United States and an additional European meeting in Paris. While the funding was relatively modest, this type of network support from ESRC enabled significant additionality and more intense interaction than would have been otherwise possible. Second, we would like to acknowledge the enthusiasm and commitment of the network members over the four-year period. Initially focused on international membership in South Africa, India, China, Australia and the US, the network was extended with members from France, Germany and Sweden. All of the members had to contribute at their own cost and time, and also made significant contributions through organizing network meetings in their own context. And we are pleased to have made new professional relationships and friends as well cementing previous ones. Special thanks to Durham University, Sheffield University, City University of Hong Kong and LATTS (Laboratoire Techniques, Territoires et Sociétés) at Université Paris Est for hosting the various network meetings. Thirdly, while primarily a research network, we would also like to thank the various policy-makers in Hong Kong, Paris and Chicago who hosted visits from the network and willingly answered our many questions. Particular thanks to Steve Gawler of ICLEI Local Governments for Sustainability, who was able to attend a number of network meetings. Fourthly, thanks to the American Association of Geographers and the Urbanization and Global Environmental

Change project for allowing us to organize meetings of the network as workshop sessions within their own international conferences. This enabled us to continue our discussions between our annual meetings. Fifthly thanks to Andrew Mould for continuing to support the production of accessible paperback books on urban studies and Egle Zigaite for taking the book through the production process. Finally, thanks to our colleagues for their support and to our families for the absences involved in developing international collaborations.

The Editors

1

INTRODUCTION

Andrés Luque-Ayala, Simon Marvin and Harriet Bulkeley

Current societies face unprecedented risks and challenges resulting from climate change. Addressing them will require fundamental transformations in the infrastructures that sustain everyday life, from energy and water provision to waste collection and mobility. Cities – the world's key infrastructural nodes securing service provision and home to the majority of the world's population – are critical in this transition. While they concentrate a range of social and economic activities that produce GHG (greenhouse gas) emissions, they are also increasingly recognized as sources of opportunities for the implementation of actions towards climate mitigation. As illustrated by the negotiations at the 2015 United Nations Climate Change Conference (COP21) and its resulting Paris Agreement, whether, how and why low carbon transitions in urban systems take place will be decisive for the success of global mitigation efforts.

Climate change increasingly features as a critical issue in urbanization policies and in the management of urban infrastructure. Research indicates that cities across the world are now engaging in strategic efforts to effect a 'low carbon transition' (Betsill and Bulkeley, 2007; Bulkeley et al., 2011; Rosenzweig et al., 2010) – to drastically cut GHG emissions while enhancing resilience and securing access to key resources. The resulting efforts appear to be making headway towards a set of policies, governing arrangements and technologies that promote a low carbon society. Yet significant questions remain unexplored. First, limited research has been undertaken internationally to comparatively examine how different cities in the North and South are responding to the challenges of climate change. Second, the balance between policy vs material interventions has not been examined, a question that is particularly relevant as many cities are failing to achieve GHG emissions reductions in spite of novel governance frameworks designed for this purpose. Finally, it is not clear whether the strategic intent of low carbon transitions can be realized in different urban contexts, nor is the extent to which advancing low carbon is transforming the political configuration of cities.

Rethinking Urban Transitions: Politics in the Low Carbon City informs a wider societal and policy debate about the potential and limits of low carbon urbanism. The book reflects critically on a decade's work (2005–2015) undertaken by a wide range of scholars across multiple disciplines – innovation studies, urban studies and environmental studies – concerned with analysing (and in some cases shaping) urban transitions. The book updates previous scholarly research on cities and low carbon transitions, actualizing the nature of the challenges and pointing specifically to the decisively political nature of such socio-technical transformations in urban systems. This involves not simply examining implementation and governance pathways, but also looking at cases of failure, retreat, contestation and the potentially regressive social and economic lock-ins established by mainstream low carbon responses. Insights from the fields of urban studies, geography, political science, history and technological transitions are combined to examine how, why, and with what implications cities bring about low carbon transitions.

The main premise of the book is that a second generation of studies on *Cities and Low Carbon Transitions* (Bulkeley et al., 2011) requires considering transitions in their political, geographical and developmental contexts. A critical reading of the significant contribution of what could be called first-generation urban transition studies (e.g. Bulkeley et al., 2011; Hodson and Marvin, 2012; Hodson et al., 2013; While et al., 2010; Geels, 2010; Rutland and Aylett, 2008 – among many others) would suggest that advancing an understanding of low carbon urbanism requires a shift from an 'extractive' model of low carbon transitions – where the focus is on reducing emissions (point-source pollution) – to an 'embedded' model of decarbonization – where low carbon logics are both rooted and disputed in and across political rationalities and development pathways. The book argues that moving towards this model requires rethinking what it means to design, practice and mobilize low carbon in the city, while also acknowledging the presence of multiple and contested developmental pathways.

Key to this shift is thinking about transitions not solely as technical, infrastructural or systemic shifts, but also as a way of thinking about collective futures, societal development and governing modes – a recognition of the political and contested nature of low carbon urbanism. *Rethinking Urban Transitions* provides a novel conceptual framework as well as empirically rich cases through which we can begin to interrogate the relevance of socio-economic, political and developmental dimensions in the making or unmaking of low carbon transitions. It also allows for a comparison of urban low carbon transition dynamics in multiple spatial and temporal contexts, as part of an in-depth examination of technological, infrastructural and policy dimensions. The book draws on a diverse range of examples (including 'world cities' and 'ordinary cities') from North America, South America, Europe, Australia, Africa, India and China, to provide evidence that expectations, aspirations and plans to undertake purposive socio-technical transitions are both emerging and encountering resistance in different urban contexts.

Governing urban low carbon transitions

An involvement of cities and municipalities in the development of responses to climate change is not new. Cities are hot spots of resources and energy consumption, by some estimates accounting for 71 to 76% of global GHG emissions from final energy use (IPCC, 2014). Large and small cities, particularly many of the so-called 'global' cities, have shown a marked strategic interest in responding to climate change. Yet collective and individual urban responses have not necessarily resulted in systematic planning efforts or in the consistent enactment of effective regulation. For over two decades, large and small cities across the world have developed initiatives and partnerships aimed at supporting global climate change mitigation efforts (cf. Betsill and Bulkeley, 2007; Bulkeley and Betsill, 2003, 2013). The focus has been largely within the domains of policy, through institutional initiatives measuring each city's contribution to climate change (via, for example, emissions inventories) while aligning various urban policies towards a set of objectives for the reduction of GHG emissions (Romero-Lankao, 2007; Dhakal, 2010; Rice, 2010; Hoornweg et al., 2011). Cities have also sought to advance climate mitigation by establishing and joining translational municipal networks and coalitions addressing energy, environmental sustainability and climate change issues (such as the International Council for Local Environmental Initiatives, Energy Cities, the C40 Cities Climate Leadership Group, and the Global Covenant of Mayors for Climate and Energy) (Betsill and Bulkeley, 2004; Acuto, 2013). Transnational municipal networks for climate change have played a critical role in positioning urban responses to climate change, while allowing cities to learn from each other and providing cities with tools for influencing policy at national and international levels (Kern and Bulkeley, 2009; Gore, 2010; Betsill and Bulkeley, 2006). The scope, actions and influence of these networks point to the scalar and multilevel nature of climate governance, subverting traditional top–down governance forms and foregrounding the horizontal and multi-stakeholder nature of urban climate governance.

A focus on policy development (e.g. via decarbonization or mitigation action plans) needs to be balanced with an acute understanding of the limitations experienced by such policies in the context of existing social and material realities of the city (Lovell et al., 2009). Considering the urban brings attention to large- and small-scale metropolitan infrastructure systems, positioning urban networks of energy, water, waste, transport, ICT (information and communications technology) and others as potential sites of intervention towards effective climate responses. This means advancing a governance of climate mitigation that acknowledges the materiality of the urban, recognizing that physical infrastructures define a great deal of how climate change is experienced and addressed. Networked infrastructures play a vital role in structuring possibilities for a low carbon urban transition, operating as both key catalysts for environmental problems and the critical means through which the governing of climate change takes place (Bulkeley et al., 2011; Rutland and Aylett, 2008). Yet, rolling out effective infrastructural responses requires transcending a purely technological approach, emphasizing the need for novel governance arrangements, an acknowledgement of a multiplicity of (human and non-human) agencies, and recognizing the social and political nature of the city's infrastructures.

Transcending institutional readings of climate governance, recent scholarly work has sought to engage with how governing low carbon in the city is accomplished by a range of state and non-state actors through their social and technical practices – a form of *urban experimentation* that often bypasses traditional funding and planning mechanisms while at the same time, in the absence of formal policy channels, creating new forms of intervention (Bulkeley et al., 2015; Bulkeley et al., 2012; Evans et al., 2016). Beyond the work of policy documents and commitments (such as local climate action plans and GHG reduction targets), responding to climate change in the city is occurring through a growing patchwork of relatively small- and medium-scale projects and interventions by public and private stakeholders alike. With these, local authorities and other urban stakeholders seek to take advantage of funding opportunities, potential strategic partnerships, or reframe local concerns in the context of the global climate change agenda – an agenda that appears to have widespread traction and political appeal (Bulkeley and Betsill, 2013). This urban low carbon experimentation "create[s] new forms of political space within the city, as public and private authority blur, and are primarily enacted through forms of technical intervention in infrastructure networks, drawing attention to the importance of such sites in urban climate politics" (Bulkeley and Castán Broto, 2013, p. 361).

Critically, governing urban low carbon transitions demands not only thinking about what cities need to do to achieve these aims, but also reflecting on what they need to *stop* doing and what they need to *undo* – the needed changes in decades- and in many cases centuries-old systems and practices supporting urban living (Bulkeley, 2015). "Un-locking carbon is as much of a challenge as locking in new and renewable forms of energy or alternative patterns of consumption" (Bulkeley, 2015, p. 1408). Moving forward towards a low carbon city deems it important to ask questions around the (often energy-intensive) infrastructures that are in place, the planning systems that shape our cities, the consumption practices that keep our economies afloat and the unbundled utility configurations of the liberalized city and what this fragmented and market-based utility landscape means for low carbon.

Materialities, intermediation and subjectivities in the low carbon city

Taking these debates one step further, this book suggest that examining urban low carbon transitions requires engaging with the various socio-materialities (technologies and infrastructures), forms of governance and intermediation, and communities and subjectivities involved in the low carbon experimentation. The book starts with an analytical framework for urban low carbon transitions (Chapter 2) that approaches low carbon transitions as the result of multiple and disparate efforts beyond simply measuring and mitigating GHGs, reinterpreting the low carbon transition as a matter of development modes. It examines the multiple competing ways of *designing, practising* and *mobilizing* low carbon urbanism, foregrounding, among other themes, multiplicity in ways of thinking about low carbon, the various agents and subjectivities involved, the many objects and flows enrolled, and the different ways in which learning and scaling up is imagined and discussed. In doing this, the framework outlines conceptual tools for asking what it means to be low carbon, what and who is involved in the transition, how is this transition likely to unfold, and how will we recognize a transition when we see it. The remaining chapters of the book are structured in three parts, dedicated to conceptual and empirical analyses examining the *technologies, materialities and infrastructures* of the urban low carbon transition (Part I), the various ways of *intermediation and governance* arrangements at play (Part II), and finally the ways in which *communities and subjectivities* become central in the making of low carbon urbanism (Part III).

Part I: Technologies, materialities, infrastructures

The first part of the book examines the objects and flows of the material world involved in the production of carbon, and the set of mechanisms and techniques that operate as material, framing and discursive devices capable of influencing both agents and objects. Recent scholarship has demonstrated that thinking about low carbon urbanism requires looking beyond policy interventions and institutional configurations, demanding an examination of the very material networks that make the city and the socio-technical dimensions of the technologies and infrastructures mobilized to bring about change (Rutherford and Coutard, 2014; Moss, 2014; Bulkeley et al., 2015). Drawing on these approaches, and foregrounding the political agency embedded within material technologies, the four chapters of Part I depict a variety of techno-material entry points to understanding both low carbon and energy transitions in Paris, Berlin and Hong Kong, Manchester and a selection of North American cities.

In Chapter 3, Jonathan Rutherford looks at the interplay between the materiality of infrastructures and the politics of low carbon. As Paris sets out to decarbonize its energy provision, Rutherford examines the material constraints to low carbon imposed by city's energy infrastructures, and the ways in which the very materiality of these infrastructures generates lock-ins and techno-political contestations to the city's low carbon objectives. Empirically, the chapter looks at the possibilities and impossibilities for expanding Paris's district heating system, the tensions between private and public interests emanating from the unclear ownership of the city's electricity distribution networks, and the city's experimentation with smart electricity as a means for dealing with carbon-intensive peak loads. It argues for an understanding of agency for low carbon as "situated in, around and between particular matters and materialities." In doing this, the chapter foregrounds the need to consider both human and material agencies, and to understand the ways in which such non-human agencies become embedded in political processes.

Chapter 4 examines the temporal dimensions of energy and low carbon transitions, focusing on the infrastructural lock-ins emerging in the political history of Berlin and Hong Kong's energy systems. Here, Timothy Moss and Maria

Francesch-Huidobro call for an analysis of historical legacies of energy production, and the extent to which these influence future low carbon options. The empirical focus is the twentieth-century need of both cities – isolated geopolitically and forcefully disconnected from wider regional, national and transnational structures – to maximize energy autarky for security purposes. Following from the 1990–97 reunification processes, the legacy of the Berlin's energy autarky creates challenges for responding to climate change and tensions between environmental sustainability, market competition and energy security.

Chapter 5 looks at the recent history of Greater Manchester, focusing on the process of envisioning a low carbon future as a technology for governing the city. Specifically, the chapter looks at how visions for a low carbon city-region rose and declined through 2006 and 2017, as the city negotiated and implemented decentralization efforts aimed at repositioning itself from an industrial city to a post-industrial entrepreneurial city. The analysis provided by Mike Hodson, Simon Marvin and Andy McMeekin suggests that low carbon visions are often highly mutable, transient and politically produced, largely in response to broader political and economic processes, priorities and power struggles. Examining the interplay between visions of the city and infrastructural priorities, Hodson, Marvin and McMeekin reflect on the implications and limitations of building low carbon capabilities through primarily economic priorities.

The final chapter of Part I also explores the process of implementing low carbon visions through infrastructural interventions. Laura Tozer, in Chapter 6, interrogates the governance documents of nine local authorities in North America to illustrate variation and difference within discourses on 'carbon neutrality'. The chapter focuses on the Carbon Neutral Cities Alliance, a transnational network of cities committed to 'deep decarbonization' – a target of at least an 80% reduction of GHG emissions by 2050. Tozer's findings show that there is significant diversity among ways of thinking about low carbon, yet there is a prevalence of hegemonic ideas about the infrastructural shape and form of such carbon neutral urban futures.

Part II: Intermediation and governance

Part II looks at a broad range of activities and structures that make up climate responses at the local level, with a specific focus on municipal governance arrangements and intermediation practices. This includes a set of institutional architectures at national, regional and local levels, operating through a disparate array of alliances, management practices, forms of finance provision, service delivery systems, forms of consulting, knowledge generation, coordination, technology provision, advocacy, lobbying, awareness-raising, dissemination and so on (Hodson and Marvin, 2012; Hodson et al., 2013). While formal institutional architectures – an important focus of the first generation of studies on urban low carbon transition (cf. Corfee-Morlot et al., 2011; Gustavsson et al., 2009) – play an important role in establishing the level of ambition and formal pathways for local climate policy, a range of formal and informal forms of organizational and technical intermediation operate as brokers for action, connecting, translating and facilitating flows of knowledge and other resources. In practice, these heterogeneous agencies articulate the options and demands of the different agents involved, supporting the alignment of visions, objectives and possibilities towards common goals (Hodson and Marvin, 2009).

The four chapters that make up Part II advance novel conceptual and empirical ways of examining the role of local governments and a variety of forms of local intermediation in the making of urban low carbon transitions. They are based on in-depth empirical cases from Sweden, India, the state of Victoria in Australia, and African cities. In Chapter 7, Susie Moloney and Ralph Horne explore inter-mediation from a relational perspective – as a set of relations that variously join, reinforce and unjoin connections that carry meanings, materials and knowledge about the making of urban low carbon. Looking at the work of a range of Climate Change Alliances in Victoria – voluntary agreements between groups of adjoining local governments sharing practices and knowledge around low carbon projects – the chapter examines the tensions that these intermediary organizations experience as both strategists and activists for low carbon. In the specific context of Australia, where climate change policy is contested, the chapter argues that local inter-mediation plays a critical role by enabling experimentation with low carbon tech-nologies and initiatives, building capacity and fostering learning across local governments, and involving a range of stakeholders towards scaling up responses.

In Chapter 8, Mikael Granberg looks at the intermediary role of local authorities in the making of a market-based low carbon urbanism. The chapter looks at the ambitious GHG reduction targets of Örebro, in Sweden, where the municipality has committed to a form of 'climate negative impact'. These aims are largely to be achieved through a green investment fund that facilitates public and private investments in renewable energy via green bonds. In this setting, the local govern-ment, operating as a hybrid boundary organization, intermediates "by trying to facilitate flows of both experience and capital between public and private actors" and also by "facilitating connections between local government and market actors." The chapter focuses on the development of capacity for low carbon urbanism at the interface between the local state and financial markets, and closes by asking questions about the potential impact of these strategies in political and democratic control within local climate action.

The final two chapters of Part II look specifically at intermediation and governance capacity for low carbon in the context of cities in the global South. Examining urban energy transitions in sub-Saharan African towns and cities, Jonathan Silver and Simon Marvin propose, in Chapter 9, the use of *urban transitions analysis* as a framework for understanding energy networks beyond the largely integrated systems across the global North. The chapter engages with the specificity of energy systems in many cities in the global South, such as a significant reliance on informal prac-tices and an infrastructural configuration characterized by limited standardization and integration. Under such conditions, the authors point to the contested politics inherent in the governing of energy networks in sub-Saharan Africa, and the need for detailed work for uncovering the agencies of various intermediaries involved in both the provision of energy and the possibility of making this into low carbon energy. The chapter also calls for an acknowledgement of "the limits of urban government and governance ... capacity and knowledge to understand and reshape urban energy systems around low carbon concerns," and through this, draws attention to the inevitably political tensions between different social, ecological and economic outcomes of low carbon pathways.

Governance limitations for advancing a low carbon urbanism are also characteristic of Indian cities. In Chapter 10, Neha Sami explains the complex and at times muddled environmental governance system present at regional and local levels in India. Based on case studies from Bangalore and Chennai, the chapter maps the institutional architecture that brings concerns around climate mitigation to the local level. In a national context where environmental governance is permanently reframed through the priorities of development and economic growth, and where climate interventions have historically been framed by narratives around co-benefits, systemic and coordinated capacity for local environmental and climate action is significantly absent.

Part III: Communities and subjectivities

The third and final section of the book focuses on the role of individuals and communities within the low carbon urban transition. These, as a range of subjectivities that are constituted in the process of governing carbon, are seen as subjects of governing as much as enrolled in practices of self-governing. In this way, Part III discusses the (human) individual and collective agencies involved in the reconfiguration of the city towards low carbon logics and forms of operation, from the agencies of neighbourhood dwellers – and their collective ways of thinking – to the agencies of workers and householders – and the efforts to bring about specific behaviours in them. The chapters included here look at the increasing institutional focus on conducting carbon conducts, as well as the collective processes of developing and enacting shared low carbon identities. They highlight how different communities of practice attempt to transform meanings and forms of relating to the city and its infrastructures in order to reduce their emissions. Through this, these chapters bring about new openings for how we understand low carbon urbanism. It is composed of three chapters looking at the experiences of various Australian cities, Ottawa and São Paulo.

In Chapter 11, Robyn Dowling, Pauline McGuirk and Harriet Bulkeley report on the organization of urban low carbon behaviour change initiatives in Australian cities, looking at a range of initiatives across local government, private sector and/ or NGO (non-governmental organization) actors. The chapter engages a Foucauldian governmentality framework to consider how behaviour change initiatives relate to the practice of low carbon transition. It thinks through the practices and dynamics of initiatives that move beyond working on the mind (i.e. an attempt to fashion a rational carbon-conscious subject through education) to modes of working on the self via material engagements, constituting a materially enrolled subject. This latter, a more innovative and possibly more effective mode of engaging subjectivities in the making of low carbon cities, fashions the reflexive, self-disciplining subject and works through materiality in shaping the conduct of low carbon subjects.

Chapter 12 focuses on cultural conflicts over 'urban intensification' – the increasingly popular trend in North American cities of increasing urban density via infill development and high-rise apartment buildings. Intensification, as a way of tackling urban sprawl, promotes a spatial configuration leading to lower GHG emissions. In this chapter, Matthew Paterson and Merissa Mueller argue that the conflicts over high-density urban projects are highly revealing of questions of agency and subjectivity in low carbon transitions, as intensification confronts prevailing high carbon cultural norms associated with more traditional car-oriented development. Using a case study of Ottawa, Canada, Paterson and Mueller reveal how attempts to transform traditionally low density and car-dependent cities encounter opposition by local communities. The chapter reveals how political conflicts around the future of the city are expressed through fundamental ambivalences, where, in spite of local support for 'liveable cities' and addressing climate change, uneasiness about changes in the character and heritage of neighbourhoods stands in the way of a set of interventions that would favour low carbon urbanism.

The final chapter of Part III, Chapter 13, revisits the experience of global South cities in order to ask what a low carbon urbanism would look like in a post-development world. Making explicit reference to the post-development debate – a critique of development as a discursive formation with political aims, a denaturalization of development as the only possible path for Asia, Africa and Latin America, and the encouragement of narratives and representations that transcend development (Escobar, 1995, 2012) – this chapter sets out a research agenda around the mobilization of low carbon (narratives and initiatives) outside dominant understandings of development. Drawing on a case study of community groups adopting DIY (do-it-yourself) solar hot-water systems in São Paulo, Brazil, Andrés Luque-Ayala reflects on the possible contribution of post-development to contemporary debates on low carbon urbanism. The chapter asks questions around the potential, limitations and implications of addressing climate change in cities in the global South via initiatives and social movements that operate through a post-development logic: social processes that take steps towards displacing liberal ontologies and capitalist imaginaries from their central role as organizing principles of both economy and society.

The book's conclusions, in Chapter 14, reflect on the specific contribution of each chapter to what we call here a second generation of studies on urban low carbon transitions, and outlines an emerging research agenda for the future.

★ ★ ★

In this book, we sought to record and analyse what is happening on the ground in cities as they set out to embrace a low carbon transition. While there are grounds for hope – in that cities worldwide are increasingly more active in developing climate responses at the local level and more powerful and influential within global climate negotiations – an in-depth examination of what and how cities have been able to embrace and advance low carbon modes of urban development reveals multiple shortcomings, barriers, power struggles and contestations. We also sought to contribute to the scholarly research on cities and climate change, by advancing novel conceptual frameworks that provide us with more sophisticated tools for understanding progress and possibilities towards a low carbon transition. We believe that, by simultaneously considering the materiality of the city, governance and intermediation arrangements, and the role of communities and subjectivities, a new conceptual understanding of the low carbon transition emerges; one where a low carbon transition is not a technical endeavour, nor a matter of systemic change, but primarily a matter of social and political debate and negotiations. This way of analysing the urban low carbon transition allows us to ask – and find multiple answers about – what it means to be low carbon, what and who is involved in the transition, how is the transition likely to unfold and how will we recognize it when we see it.

Most importantly, we set out to identify different pathways for the city to become low carbon, decentring the central tenet of lowering GHG emissions and foregrounding ways of thinking about development and collective futures. The low carbon transition is not a singular problem, but rather a problem full of multiplicities, meanings and potential responses. As cities increasingly engage with efforts towards climate mitigation, it becomes clear that "the economic, social and political consequences of such transitions are variously figured." For some, the low carbon transition is an opportunity "for the development of a green economy and clean technologies, while for others it represents the need to shift to radically new economic systems which are more localised in their forms of production and consumption" (Bulkeley, 2015, p. 1406). As we have seen in over two decades of action and research at the interface between cities and climate change, it is a problem that is continually in transformation, from one of environmental protection and decarbonization to one of green growth, economic resilience, development logics and more. This continuous rethinking of the urban low carbon problem is not so much a function of technological progress, but the result of political challenges and contestations. Acknowledging multiplicity and contestation in the design, practice and mobilization of the low carbon city is likely to better equip us for both researching and advocating for the much needed environmental, societal and political transformations of the contemporary world – of which becoming low carbon represents only a small one, yet provides us with a big opportunity for bringing about societal change.

References

Acuto, M. (2013) City leadership in global governance. *Global Governance*, 19(3), 481–498.
Betsill, M. and Bulkeley, H. (2004) Transnational networks and global environmental governance: the cities for climate protection program. *International Studies Quarterly*, 48(2), 471–493.

Betsill, M. and Bulkeley, H. (2006) Cities and the multilevel governance of global climate change. *Global Governance*, 12(2), 141–159.

Betsill, M. and Bulkeley, H. (2007) Looking back and thinking ahead: a decade of cities and climate change research. *Local Environment*, 12(5), 447–456.

Bulkeley, H. (2015) Can cities realise their climate potential? Reflections on COP21 Paris and beyond. *Local Environment*, 20(11), 1405–1409.

Bulkeley, H. and Betsill, M. (2003) *Cities and Climate Change: Urban Sustainability and Global Environmental Governance*. London: Routledge.

Bulkeley, H. and Betsill, M. (2013) Revisiting the urban politics of climate change. *Environmental Politics*, 22(1), 136–154.

Bulkeley, H. and Castán Broto, V. (2013) Government by experiment? Global cities and the governing of climate change. *Transactions of the Institute of British Geographers*, 38(3), 361–375.

Bulkeley, H., Castán Broto, V. and Edwards, G. (2015) *An Urban Politics of Climate Change: Experimentation and the Governing of Socio-technical Transitions*. New York: Routledge.

Bulkeley, H., Castán Broto, V., Hodson, M. and Marvin, S. (2011) *Cities and Low Carbon Transitions*. London: Routledge.

Bulkeley, H., Hoffmann, M.J., VanDeveer, S.D. and Milledge, V. (2012) Transnational governance experiments. In: Biermann, F. and Pattberg, P. (eds.) *Global Environmental Governance Reconsidered*. Cambridge, MA: MIT Press, pp. 149–171.

Corfee-Morlot, J., Cochran, I., Hallegatte, S. and Teasdale, P.-J. (2011) Multilevel risk governance and urban adaptation policy. *Climatic Change*, 104(1), 169–197.

Dhakal, S. (2010) GHG emissions from urbanization and opportunities for urban carbon mitigation. *Current Opinion in Environmental Sustainability*, 2(4), 277–283.

Escobar, A. (1995) *Encountering Development: The Making and Unmaking of the Third World*. Princeton, NJ: Princeton University Press.

Escobar, A. (2012) *Encountering Development: The Making and Unmaking of the Third World*. 2nd ed. Princeton and Oxford: Princeton University Press.

Evans, J., Karvonen, A. and Raven, R. (2016) *The Experimental City*. London: Routledge.

Geels, F. (2010) The role of cities in technological transitions. In: Bulkeley, H., Castán Broto, V., Hodson, M. and Marvin, S. (eds.) *Cities and Low Carbon Transitions*. London: Routledge, pp. 13–27.

Gore, C.D. (2010) The limits and opportunities of networks: municipalities and Canadian climate change policy. *Review of Policy Research*, 27(1), 27–46.

Gustavsson, E., Elander, I. and Lundmark, M. (2009) Multilevel governance, networking cities, and the geography of climate-change mitigation: two Swedish examples. *Environment and Planning C: Government and Policy*, 27(1), 59.

Hodson, M. and Marvin, S. (2009) Cities mediating technological transitions: understanding visions, intermediation and consequences. *Technology Analysis and Strategic Management*, 21(4), 515–534.

Hodson, M. and Marvin, S. (2012) Mediating low-carbon urban transitions? Forms of organization, knowledge and action. *European Planning Studies*, 20(3), 421–439.

Hodson, M., Marvin, S. and Bulkeley, H. (2013) The intermediary organisation of low carbon cities: a comparative analysis of transitions in Greater London and Greater Manchester. *Urban Studies*, 50(7), 1403–1422.

Hoornweg, D., Sugar, L. and Gomez, C.L.T. (2011) Cities and greenhouse gas emissions: moving forward. *Environment and Urbanization*, 23(1), 207–227.

IPCC (2014) *Climate Change 2014: Mitigation of Climate Change. Contribution of Working Group III to the Fifth Assessment Report of the Intergovernmental Panel on Climate Change*. Cambridge: Cambridge University Press.

Kern, K. and Bulkeley, H. (2009) Cities, Europeanization and multi-level governance: governing climate change through transnational municipal networks. *Journal of Common Market Studies*, 47(2), 309–332.

Lovell, H., Bulkeley, H. and Owens, S. (2009) Converging agendas? Energy and climate change policies in the UK. *Environment and Planning C: Government & Policy*, 27(1), 90–109.

Moss, T. (2014) Socio-technical change and the politics of urban infrastructure: managing energy in Berlin between dictatorship and democracy. *Urban Studies*, 51(7), 1432–1448.

Rice, J.L. (2010) Climate, carbon, and territory: greenhouse gas mitigation in Seattle, Washington. *Annals of the Association of American Geographers*, 100(4), 929–937.

Romero-Lankao, P. (2007) How do local governments in Mexico City manage global warming? *Local Environment*, 12(5), 519–535.

Rosenzweig, C., Solecki, W., Hammer, S.A. and Mehrotra, S. (2010) Cities lead the way in climate-change action. *Nature*, 467(7318), 909–911.

Rutherford, J. and Coutard, O. (2014) Urban energy transitions: places, processes and politics of socio-technical change. *Urban Studies*, 51(7), 1353–1377.

Rutland, T. and Aylett, A. (2008) The work of policy: actor networks, governmentality, and local action on climate change in Portland, Oregon. *Environment and Planning D: Society and Space*, 26(4), 627–646.

While, A., Jonas, A. and Gibbs, D. (2010) From sustainable development to carbon control: eco-state restructuring and the politics of urban and regional development. *Transactions of the British Institute of Geographers*, 35(1), 76–93.

2

RETHINKING URBAN TRANSITIONS

An analytical framework

Andrés Luque-Ayala, Harriet Bulkeley and Simon Marvin

1 Introduction

Over the course of the past two decades cities have become critical stakeholders in the development of responses to climate change. With urban populations growing rapidly across the world, from a total of 30% in 1950 to 54% in 2014 and an estimated 66% by 2050 (UN Population Division, 2014), how cities engage with efforts to mitigate climate change is of paramount importance for the governance of climate change. Acting as pioneering and visionary leaders in the international arena, cities and local authorities worldwide are leading efforts to translate global climate needs and commitments into meaningful action on the ground.

Since the mid-2000s a range of scholarly work has been tracking cities' involvement in climate action and their efforts towards GHG (greenhouse gas) reduction. From evaluating targets and commitments within local policy, to tracking the collective efforts of transnational networks of cities and examining the transformation of urban infrastructures into low carbon systems, this scholarly work has laid out invaluable foundations for an analysis of urban low carbon transitions. Building on this work, and developing a sympathetic yet critical analysis of what we refer to here as first-generation studies on low carbon urbanism, this chapter provides a broad framework for taking steps towards a second-generation of studies on low carbon urban transitions – one where low carbon transitions are interpreted primarily as social, political and developmental processes.

The chapter argues that a next generation of urban transition studies needs to be more forthcoming in considering transitions in their political and geographical context. A critical reading of the significant contribution of the first-generation studies on low carbon transitions would suggest that advancing an understanding of low carbon urbanism requires a shift from an 'extractive' model of transitions, where the focus is on reducing emissions (an end-of-pipe approach), to an 'embedded' model of decarbonization that transcends narratives around systemic change and reconceptualizes low carbon as a matter of development modes. Key to this shift is thinking about low carbon transitions not solely as technical or infrastructural shifts but also as a way of thinking about society, its politics and economic processes, and its ways of envisioning the development of collective futures.

The proposed framework examines the multiple and competing ways of *designing, practising* and *mobilizing* low carbon urbanism – three pivotal dimensions in the making of the low carbon city. Foregrounding the politics of low carbon, the framework asks what does it mean to be low carbon, what and who is involved in the transition, how does the transition unfold and how will we recognize one when we see it. Its purpose is threefold. First, it provides a conceptual lens through which we can begin to interrogate and compare urban low carbon transitions in multiple contexts, framed by a set of concepts that cut across empirical and theoretical domains. Second, it reflects critically on over a decade's work, between 2005 and 2017, undertaken by a wide range of scholars across multiple disciplines – urban studies, geography, environmental studies and innovation studies, among others – concerned with analysing (and in some cases shaping) urban transitions. This is not an attempt to provide a comprehensive review of the literature. Rather, it is a summary review of pivotal positions and works with the aim of sympathetically and constructively evaluating key contributions, identifying key elements for an emerging research agenda, and distinguishing potential conceptualizations that might inform a second generation of low carbon transition studies. Finally, the framework is designed to be informing of wider societal and policy debates about the potential and limits of low carbon urbanism, hinting at the strategic direction of future policy responses. In doing this the framework operates at both analytical and normative levels, opening a space for critical thinking while also signalling a way forward.

The framework, while articulated by the editors, is the result of a collective process of reflection carried out with and alongside the authors participating in the development of the individual chapters of the book as well as some practitioners who joined us at key moments. It developed slowly, over the course of four years, through individual and group meetings, academic workshops, discussions and visits to low carbon initiatives in several of the cities where each author conducts research. As a result, the framework embraces plurality and multiplicity from the outset, having been informed by an extensive range of disciplinary backgrounds, geographical locations (and their contexts), research interests and even academic cultures and languages. The site visits to 'exemplary' and 'mundane' urban low carbon initiatives, characteristic of each meeting and workshop, were pivotal in grounding the ideas discussed, allowing for a cross fertilization between theory and empirics. The framework purposefully draws from distinct and diverse theoretical positions, namely theories of governmentality, political economy and urban political ecology, social practice theory and institutional approaches. The rationale behind this broad (and perhaps, at specific junctions, disparate) combination of conceptual entry points is not to provide an integrative device, but rather one that

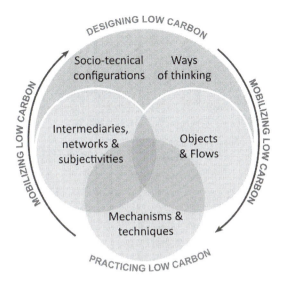

FIGURE 2.1 Towards a second-generation studies of urban low carbon transitions: the threefold framework.

'sets the frame' for analysis, allowing research on urban low carbon transitions to take different analytical pathways.

The rest of the chapter is structured into the three dimensions constituting the framework (Figure 2.1). First, it focuses on *designing low carbon* urbanism, looking at the relevance of unpacking ways of thinking about low carbon in the city and the emerging socio-technical configurations associated with this process. Second, it centres on *practising low carbon urbanism*, locating the analytical gaze on the agencies and subjectivities, objects and flows, and mechanisms and techniques at stake. Third, the chapter concentrates on *mobilizing low carbon* urbanism, providing insights for an analysis of how transitions unfold. The conclusion provides a synthesis of first- and second-generation studies on urban low carbon transitions, advocating for an acknowledgement of multiple junctions towards the low carbon city and embracing plurality in the pathways through which a low carbon urbanism might occur.

2 Designing low carbon urbanism

Examining and advancing emerging forms of low carbon urbanism demands starting with a simple question: *what does it mean to be low carbon?* For many, answering this question takes us a step closer to a techno-utopian clean, smart, sustainable and green future – a time ahead purposefully designed to be low carbon. For others it is a moot question; one with no single answer, as the multiple voices and presences that make society are inevitably bound to disagree in the search for a unified and hegemonic single vision of the future. However, to ask the question, as an entry point for rethinking low carbon transitions, already signals the extent to which a low carbon transition is a matter of societal development: a political process that deems it important to acknowledge multiple voices, a shared belief in some form of common future, and collective action. Whose voices are acknowledged, which common futures are believed in, and what collective actions mobilized, are all matters of 'design' – a combination of socio-technical configurations and ways of thinking about society and its dependence on fossil fuels and other processes that generate GHGs.

While recognizing the multiplicity and contradiction embedded in current attempts to realize low carbon, here we propose a way of analysing low carbon urbanism based on the acknowledgement and development of novel ways of thinking about the low carbon–development interface. Decarbonization is then not confined to setting targets to reduce GHG emissions, but involves a host of more or less explicit ways in which carbon comes to be problematized and acted upon in relation to, our particular concern, the city. Advancing a low carbon configuration for the urban is, as a result, potentially a transformative process. The design of low carbon urbanism involves establishing a shared understanding of the challenges of achieving low carbon in relation to particular urban contexts of intervention. Such understanding powerfully shapes the identification of social and technological interventions to be trialled, the form of coalitions and selection of intermediary institutions, and the collective agreement on the governance principles to be followed. There are two dimensions at play in unpacking the design of low carbon urbanism. The first is an analysis of socio-technical configurations, focusing not so much on the intervention itself, we argue, but on how the existing context shapes the spaces for innovation around low carbon urbanism. The second is an analysis of the ways of thinking about the relations between low carbon and urban development priorities, an inquiry that seeks to reinsert multiple geographies and contextually embedded issues into urban transitions.

Examining the emerging socio-technical configurations

The existing context of an intervention shapes its possibilities. In the case of low carbon urbanism, this context provides an opening for opportunities as much as the lock-ins that impose resistance. Considering the extent to which a broad context shapes possibilities for transformation within sustainability transitions is not new. Research on socio-technical transitions (STT) and the multilevel perspective (MLP), for example, when examining the role that technological niches play in advancing sustainability innovations, recognizes the extent to which contextual landscapes apply pressures over existing and established regimes (Smith et al. 2010). Smith et al. (2005), in discussing the use of STT and MLP approaches in the context of transitions to sustainability, illustrate the notions of *niche, regime* and *landscape* – three foundational notions within STT and MLP literatures – through an example of electricity generation. In this example the electricity-generating regime is made of, among others, large-scale centralized power plants (such as coal- or gas-fired, nuclear-based or hydro-powered) working in tandem with high voltage grid infrastructures, alongside the rules and practices associated with the functioning of these systems. Novel technologies such as wind turbines, photovoltaic installations or other renewable technologies, and associated practices, constitute an emerging innovation niche that poses a challenge to the dominance of the regime. Such challenge plays out in the context of wider social and political landscapes that put pressure on regimes, such as the liberalization of energy markets, public debates around the risks of nuclear technologies or the benefits of a 'hydrogen economy', or cultural trends around consumption. Landscapes, sociotechnical, cultural and political in nature, create windows of opportunity towards positive responses (Smith et al., 2005; Berkhout et al., 2004). In foregrounding the role of context in the making of socio-technical configurations for low carbon urbanism, the various case studies included in this book illustrate how contextual configurations define the right set of circumstances as much as the carbon lock-ins that dominate the regime. Context, therefore, should not be seen through an evolutionary and unilinear lens where the landscape's role is limited to triggering positive adaptive responses.

Technological innovation is commonly seen as a first and required step in developing a low carbon future. In this context, for almost two decades, scholars advancing STT and MLP approaches have developed a powerful and valuable conceptual language for understanding sustainability transitions across a variety of scales. STT and MLP scholars have pointed to the extent to which existing systems are difficult to dislodge given processes of stabilization and lock-in developed by incumbent regimes. This leads to path dependency and 'entrapment', two conditions that constrain the emergence of alternatives (Grin et al., 2010). A key means through which the governance of transitions is thought to proceed is through strategic niche management (Kemp et al. 1998) – whereby governments, or other actors, deliberately seek to establish conditions under which niches for sustainable innovation can grow and 'break through' existing regime conditions.

MLP advocates for an analysis of the role of technological niches and experiments in the making of more sustainable systems. Operating within relatively stable regimes, these niches play an important role in advancing low carbon and sustainability transitions (Schot and Geels, 2008). Niches function as protected environments that provide space for the development, testing and failure of novel innovations, and where networks advancing more sustainable interventions can be supported and sustained (Smith and Raven, 2012). This work has made significant contributions to the conceptualization of socio-technical change and the analysis of transition dynamics and processes.

Early critiques of the MLP approach called for the development of more nuanced and revised approaches on three fronts: spatial context, agency and power. First, the MLP perspective was strongly criticized for disregarding the role of geography, time and space while drawing excessive attention to national scales (Lawhon and Murphy, 2012; Hodson and Marvin, 2010). Niche innovation was seen primarily as the result of stakeholders operating at the national level, disregarding the role of local and regional stakeholders in its development. One important implication of an undue emphasis on the national scale is a limited recognition of the role of the general public, with its contribution narrowly framed in the context of consumption patterns. The city was taken to be an empty container where innovation was likely to occur, yet the urban condition itself, and how it could influence processes of innovation, was not given enough attention (Geels, 2010). Transitions were, for the most part, a matter of the contemporary present; little attention was given to time dimensions and how history influences the possibilities for contemporary innovation (Moss, 2011). More recent iterations of transition studies and the MLP approach have incorporated a more nuanced spatial understanding (Truffer and Coenen, 2012). Out of this, it became clear that socio-technical configurations for low carbon transitions occupy a multiplicity of spaces, operating in and across multiple geographies. Urban sustainability transitions are unique not only because of the scale at which they (primarily) occur, but because they involve and intersect a range of sector specific transitions (e.g. energy, water, transport, etc.) – an insight that establishes empirical and conceptual distinctions between urban transition studies and the more traditional domain-specific studies of sustainability transitions, calling for city-specific tools of analysis (Frantzeskaki et al., 2017). They are multiscalar in nature (cf. Bulkeley and Betsill, 2005), deeply grounded in local conditions yet shaped by debates and actions at regional, national and global scales. A second generation of urban transition studies needs to build on these insights, also recognizing that transitions are not made solely of the present, and that a historical dimension is key to understanding their formation.

Second, while research has so far tended to focus on questions of the design and production of niches and experiments, analysing these processes requires that more attention is given to dynamics of agency and power – the practices of governance on the ground (Smith and Raven, 2012). Recent perspectives within STT research have introduced politics and power as variables affecting the development of niches, pointing to heterogeneity within transitions, the contested processes by which legitimacy is built, and the extent to which regimes not only take advantage of innovation but also can impose resistances associated with broader political-economic conditions (Geels, 2014; Raven et al., 2016). A new generation of urban transition studies needs to build on these insights, responding to the growing calls for engagement with the political contexts within which transitions evolve, and the processes through which niches and experiments are related to these wider systemic contexts (Coenen et al., 2012; Shove and Walker, 2007; Meadowcroft, 2007). For a second generation of transition studies it will therefore be critical to engage with the multiplicity of sites and forms of agency involved, and to account for the ways in which contestation forms an active part of transition processes.

Unpacking ways of thinking about low carbon

An important premise of this book is that examining how the very possibility of a low carbon future is conceptualized within different settings acts as a starting point to understanding the embedding of low carbon in different urban contexts. The various case studies included in this book do not aim at making sense of low carbon transitions through extracting lessons from a few ideal cases. Rather they do so by prioritizing an analysis of ways of thinking about low carbon, through unpacking (and at times foregrounding) the emerging rationalities and subjectivities created in the search for a low carbon city. Within a first generation of urban low carbon transition studies, low carbon has often been seen a matter of techno-economic innovation with the primary aim of reducing GHG emissions. Here the main task is one of either increasing the efficiency with which we use carbon (e.g. replacing some forms of inefficient technology or design with another) or substituting carbon-intensive systems and practices (such as heralded by renewable energy technologies or electric cars). As briefly discussed in the previous pages, such emphasis on techno-economic innovation prioritizes flat conceptions of power and agency, overlooking the scalar challenges and possibilities associated with multiple and overlapping, formal and informal, governing regimes. Beyond techno-economic innovation, there are multiple other ways of thinking about low carbon urbanism.

We argue that, by abandoning an exclusive focus on GHG reduction and climate mitigation, the low carbon transition is analytically repositioned as a development process; one with potentially transformative political and economic implications. Transitions, rather than being the result of clean and purposeful ways of scaling niches, are contingent and politically contested processes where a multiplicity of systems, agents and scales come together in an attempt to reconfigure social interests, political arrangements and technology. In acknowledging the political aspects of low carbon transitions, we open up possibilities for an analysis of the mobilization of low carbon towards a broad variety of social and political objectives, including forms of carbon control, urban ecological security and issues of social justice and urban inequality. While this is not an exhaustive list of the critical issues at stake, it starts to provide a range of thinking frameworks through which we may conceptualize the embedding of low carbon urbanism within particular urban contexts. Each will be discussed in turn.

In most industrialized countries, issues related to the need to control carbon have emerged as "the new 'master concept' of environmental regulation" (While et al., 2010, p. 77). Such an emerging mode of environmental governance reflects wider political concerns, including energy security, resource security, issues of oil dependency and, most importantly, the search for securing economic growth through environmental interventions. Throughout the 1980s and 1990s environmental governance prioritized forms of ecological modernization, heralding an optimistic view of technology and institutional change in solving environmental problems while justifying environmental interventions on account of their economy benefits (Hajer, 1995). However, since the 2000s, environmental governance has responded to climate change and the problematic resulting from an economy based on fossil fuels, arguing for the need to transition to a low carbon capitalism. Like ecological modernization, an environmental governance based on carbon control reflects the possibility of gaining profits from a low carbon transition. Critiques of the emerging modes of carbon control point to a preference for market-based solutions, with the state avoiding the political implications of direct taxation by passing the responsibility of carbon reduction to firms, citizens and localities. Carbon control can be seen as a political-economic fix to a crisis in capitalistic modes of production, enacted via state environmental regulation (While et al., 2010). Here the low carbon transition remakes the city into a space of carbon flows, with the tools of accountancy underpinning a new carbon calculus aimed at reconstructing the urban (While, 2014) – the low carbon city, one where environmental flows and carbon responsibilities are closely monitored, becomes a space of rational calculability where politics are determined through environmental calculations.

Another way in which low carbon urban transitions are being conceptualized is through the notion of 'urban ecological security'. This approach points to a growing trend, particularly present in large cities, to reconfigure urban infrastructures towards securing the key resources required for ecological and material reproduction – notably, those infrastructures needed to maintain and enhance economic growth (Hodson and Marvin, 2009b). This common urban response to climate change is leading to "new styles of infrastructure development that privilege spatial and socio-technical configurations of infrastructure around 'strategic protection'... [and] 'autarky'" (Hodson and Marvin, 2009b, p. 200). Reconfiguring infrastructures is a matter of urban governance, manifested in cities' efforts to increase local control over key resources ('re-internalize' and 're-localize'), aiming towards self-sufficiency by fostering closed loops and circular metabolisms, and developing protection against climate impacts (flooding, temperature change, others). As illustrated by the various energy strategies of London, Amsterdam, New York City and Melbourne of the past decade, the result is a 'metropolitanization of ecological resources', where resources and socio-technical infrastructures are "increasingly reincorporated and re-enclosed within the metropolitan boundary" (Hodson and Marvin, 2009b, p. 205). Urban responses to climate change are highly based on transnational partnerships with other cities and the development of alliances for cooperation. This strategic response of world cities is considered to have a global political and economic relevance, where cities like London, Paris, New York or Rio de Janeiro play 'exemplary' roles, with the city operating as a key site for the experimentation and demonstration of new infrastructural technologies (May and Marvin, 2009; Hodson and Marvin, 2007, 2009a; Bulkeley, 2006). The complex, and often interwoven, topological spaces of urban climate governance suggest that any straightforward relationship between local niches and national regimes needs to be reconsidered in a second generation of transition studies. Moreover, the growth of experimentation as a mode of governance suggests that niches do not hold a monopoly on experimentation. Rather, this is a mode of governing through which both incumbent and emergent actors are seeking to enact the urban politics of climate change.

More recently, low carbon interventions in cities have been conceptualized as a matter of equity and justice – through an analysis of their potential to contribute to the achievement of social justice or, alternatively, their capacity to foster conditions of exclusion and inequality (Bulkeley et al., 2013, 2014; Marino and Ribot, 2012; Fuller, 2017). This scholarship considers the extent to which the city and its engagement with the logics of low carbon provides new perspectives around issues of climate justice. In cities in the global North, for example, there is a greater tendency to position the individual as the locus of climate justice, while in the global south the emphasis lies in collective risks and benefits. This leads to significantly different conceptualizations of the urban low carbon transition. The resulting analysis points to how, when issues of justice are considered within urban low carbon interventions, notions of distributive justice dominate over procedural justice (Bulkeley et al., 2014). Engaging with issues of justice in the city enables an analysis that goes beyond the discursive mobilization of global climate justice principles and equivalences, examining both the discursive and material ways in which inequality is produced and contested at the local level, and the political economies/ecologies that shape rights and responsibilities.

3 Practising low carbon urbanism

The second dimension constituting the framework draws attention to the practice of low carbon urbanism. To think about the practices involved in the making of a low carbon city means bringing low carbon into the everyday. This perspective recognizes that the infrastructural arrangements of the city are not simply critical to the social practices that make the urban; they constitute them (Bulkeley et al., 2016). In pointing to the everyday, thinking about the practice of low carbon urbanism brings routine and mundane to the fore (Gram-Hanssen, 2011; Strengers, 2013) – a way of constituting low carbon in the everyday. Stepping away from conceptual approaches that foreground the role of the individual in allocating environmental responsibilities, a consideration of practices in the making of a low carbon city recalls both the agency of the collective and the co-constitution of the social and the infrastructural. We argue that attention to the practice of low carbon urbanism involves asking *what and who is involved in the transition?* This means working with and upon three distinct sets of elements. First, the agents and subjectivities that operate as initiators and partners or act as subjects of intervention. Second, the objects and flows of the material world involved in the production of carbon. Finally, the mechanisms and techniques that operate as material, framing and discursive devices capable of influencing both agents and objects.

Agents in practice: intermediaries, networks and subjectivities

Governing low carbon in the city transcends local and institutional arenas, involving a broad range of agents across and beyond the local state (Bulkeley and Betsill, 2005). On the one hand, non-state actors, from private to non-profit sectors, increasingly play a significant role in shaping, configuring and contesting climate responses in the city (Bulkeley and Betsill, 2013). Take for instance the work of the many local transition initiatives across the world, such as Transition Town Brixton in London or the Nelson Mandela Bay Transition Network in Port Elizabeth, South Africa, where local collectives are developing their own version of a low carbon city through local food production and sustainable transit initiatives, among others. Or the many businesses and organizations increasing their energy efficiency and reducing GHG emissions by voluntarily adopting the LEED (Leadership in Energy and Environmental Design) green building certification programme. A low carbon city is always in the making in spite of the presence or absence of municipal policies and interventions for low carbon. On the other hand, making low carbon in the city is a multiscalar process, made of interactions between agents located at different scales, from the dwelling and the city to transnational and global arenas. Looking at climate change from the lens of these diverse scales and agents challenges the traditional understanding of environmental governance that assumes that decisions are cascaded from international to national to local levels (Betsill and Bulkeley, 2007; Bulkeley, 2005; Bulkeley and Betsill, 2003; Betsill and Bulkeley, 2004). The scales and agents involved in making the low carbon city are broad and diverse. Different entities operate as initiators and partners, aiming to influence a set of agents. These can include government organizations at local, regional or national levels, as well as private sector organizations, non-profits or community groups, among others. They interact in a variety of modes of action, discussed further below, including forms of intermediation, the development of networks and the production of new subjectivities.

Understanding the low carbon transition requires detailed attention to the varied roles that different actors play, the ways in which these actors mobilize climate change logics in the context of their own agendas, and the ways in which such efforts result in contested agendas or aligned objectives in relation to the city's future. A focus on intermediation acknowledges the need to think about climate change responses beyond institutional contexts, where agents located at different governance levels (e.g. municipalities, national governments or transnational organizations) interact with both state and non-state partners (e.g. business, academia, community associations, non-governmental organizations) in the visualization and implementation of low carbon responses. Here the notion of 'intermediation' (Bourdieu, 1984; Callon, 1986; Iles and Yolles, 2002; Latour, 2005; van Lente et al., 2003) plays a key role, as multiple agents and organizations get involved in the development of capacity for low carbon, through activities such as management, finance provision, service delivery, consulting and knowledge generation, coordination, technology provision, advocacy, lobbying, awareness-raising, dissemination and others (Hodson and Marvin, 2012; Hodson et al., 2013). Intermediaries, thus, play a role as brokers, connecting, translating and facilitating flows of knowledge and other resources. Intermediary organizations, operating between different social interests (and technologies), produce outcomes that would not have been possible without their involvement (Hodson and Marvin, 2010; Hodson et al., 2013; Evans and Karvonen, 2014). These intermediation processes play a key role in translating and articulating the objectives, options and demands of different agents involved, supporting the alignment of visions, objectives and possibilities towards common goals. Yet, climate change intermediation processes are not exempt from tension and conflict, and can also reveal a multiplicity of – sometimes contradictory – urban agendas at play in the attempt to advance a low carbon city.

The 'network' stands out as a unique type of agent in the making of a low carbon city. Networks, acting both as a central node of action and an intermediary between other agents, have been pivotal in advancing low carbon responses at city and municipal levels. Transnational networks of cities working on climate, energy and environmental issues, operating in the context of global climate agreements, have played a critical role positioning cities as central stakeholders within international climate negotiations (Bulkeley, 2015). The Cities for Climate Protection programme, Energy Cities (the European Association of Local Authorities in Energy), the International Council for Local Environmental Initiatives, and more recently the C40 Cities Climate Leadership Group, the Asian Cities Climate Resilience Network or the 100 Resilient Cities programme of the Rockefeller Foundation, among others, have enabled cities to learn from each other, fostering a horizontal form of climate governance (Bulkeley and Betsill, 2003; Kern and Bulkeley, 2009; Feldman, 2012; Gore, 2010). They have enabled cities to multiply their influence, horizontally across cities as well as vertically with other levels of government (Betsill and Bulkeley, 2006). Transnational municipal networks are also perceived as a way for cities to gain new room for political manoeuvring in the development of climate action; in responding to climate change, cities have seen in these networks a possibility for advancing broader strategic objectives and interests (Heinelt and Niederhafner, 2008; Hodson and Marvin, 2009b; Hodson and Marvin, 2010; Kern and Bulkeley, 2009). Given their ability to garner widespread support and develop partnerships with a variety of stakeholders across civil society, they play an instrumental role in creating multisectoral partnerships within urban areas, including with the private sector and the third sector (Bulkeley and Schroeder, 2012; Bontenbal, 2009).

A third way in which a range of agents interact in the making of the low carbon city is through imagining and producing new identities and subjectivities. The creation of subjects that are amenable to the low carbon objective is fundamental to this process. Drawing on conceptual frameworks derived from a Foucauldian governmentality (Foucault, 2009; Dean, 2010; Legg, 2007; Murray Li, 2007; Dowling, 2010), we argue that imagining a particular type of subjects, and through this endowing them with specific powers, is of relevance for the environmental reconfiguration of the city. Arguably, governing carbon locally "operates by educating desires and configuring habits, aspirations and beliefs" (Murray Li, 2007, p. 5). This occurs through processes of subjectification, such as the generation of climate awareness and purposeful attempts to foster normative notions of citizenship alongside particular energy and environmental behaviours. Examples of this are the Climateers programme promoted by a partnership between WWF (World Wildlife Fund) and HSBC in Hong Kong – where members of the public are invited to engage with climate issues by making changes in their behaviour and calculating their contribution to GHG reduction via an online calculator (Bulkeley et al.,

2015) – and the Solar Cities Programme of India's Ministry of New and Renewable Energy – where, in promoting the programme, environmental identities collude with national identities for the mobilization of the citizen into action (Luque-Ayala, 2014). Such subjectification, as an attempt at casting those who are to be governed, is based on the mobilization of authority and power towards "creating and enrolling responsibilized carbon-relevant citizens as governable environmental subjects" (McGuirk et al., 2014, p. 138, in reference to Rice, 2010). Subject formation is undertaken not only by the state, but also by a multiplicity of other actors (Rose, 1999; Barnett et al., 2008). It is a process that targets the epistemological framing of those to be governed, reconfiguring their status, capacities, desires and agencies (Legg, 2007).

Unfolding low carbon through objects and flows

Practising low carbon also unfolds through the transformation of a series of objects and flows. A revised research framework for low carbon urban transitions needs to acknowledge the materiality of the city and the way in which physical infrastructures, flows and built environments clearly structure how climate change as an urban problem is conceived and addressed. Networked infrastructures play a vital role in structuring low carbon urban transitions. From transport systems and water treatment plants to energy generation stations, among others, these are the socio-material elements that very often stand at the heart of the low carbon transition. They are both key catalysts for environmental problems and the critical means through which the governing of climate change takes place (Bulkeley et al., 2014; Rutland and Aylett, 2008). They are both the means and the object of governing, and should not be understood exclusively as a collection of material components – the pipes, cables, buildings and flows of resources – but as socio-technical assemblages that include human and non-human agents, including regulations, rules, behaviours, standards, organizations and communities of practice and knowledge (Graham and Marvin, 2001). We argue that discourse analyses on their own, focusing exclusively on policies, targets, statements and aspirations, fail to recognize how policy-making is implemented in the context of existing material realities (Lovell et al., 2009). Analyses of climate change that do not recognize the materiality of the city, very often associated with the policy implementation stages, fail to provide an accurate account of the scale and practical impact of the transition. Such "focus on discourse as the key structuring force of policy problems underplays the material realities within which policy operates" (Lovell et al., 2009, p. 93).

It is important to note that the materiality of the city is not static; it operates through perpetual movement, signalling to a particular urban nature. Understanding the city as a process of urbanizing nature points to how urban infrastructures, rather than separating nature and the city, bring them closer together (Swyngedouw and Kaika, 2000). The flows of energy, water, waste and transport that enable life in the city are also flows of carbon – at times fossilized remnants of the earth's past natures, and at times the result of contemporary natural processes. The urban infrastructures that underpin such flows do so by mobilizing and transforming natural resources, acting as interfaces between nature and society (Monstadt, 2009) while transferring carbon across a range of environments. However, far from being distinct and clearly separated ontological entities, nature and the city operate in a mesh of "networks and flows of natural elements, social power relations and investment cycles" (Kaika, 2005, p. 5). Understanding how these technonatures are produced, and how carbon circulates through them, helps us uncover the deep physical, social, cultural, economic and political entanglement of low carbon interventions.

Embracing the relational approach outlined above forces us to examine also the very flows that are shaped and directed by the infrastructures of the city, and how the transformation of such flows under low carbon rationalities has connotations beyond the purely material. Studies of the production of technonatures in the city have uncovered the deep physical, social, cultural, economic and political entanglements present within urban infrastructures (Kaika, 2005; Kooy and Bakker, 2008; Giglioli and Swyngedouw, 2008). An understanding of how infrastructures mediate between nature and the city uncovers the urbanization of nature, not as a technological or engineering problem but as a socio-political process with implications for the wielding of power and distribution of resources (Gandy, 2004; Kaika and Swyngedouw, 2000). When linked to debates around low carbon systems as well as renewable and sustainable energies, it becomes clear that "questions of socio-environmental sustainability are therefore fundamentally political questions" (Swyngedouw, 2006, p. 119; see also McFarlane and Rutherford, 2008; Monstadt, 2009). Such political understanding of the city–nature relationship challenges aseptic and technical notions of sustainability, demanding more in-depth social analysis for any low carbon intervention.

The final consideration when thinking about the objects and flows involved in the urban low carbon transition has to do with a form of absence – the absence of the flow resulting from the moments of infrastructural breakdown, and the opportunities emerging from the disruption of the everyday. Urban infrastructures, operating in invisible, taken-for-granted and 'black-boxed' ways (Leigh-Star, 1999; Hughes, 1983), regain visibility upon breakdown. The space between breakdown and restoration positions repair and maintenance as a space for learning and transformation (Graham and Thrift, 2007). Repair and maintenance, two overlooked dimensions in the study of infrastructural systems, are not secondary or derivate but pivotal to the advancement of low carbon urbanism. In the context of urban experimentation with climate change, repair and maintenance have been identified as key dimensions through which low carbon urbanism is both made and maintained (Castán Broto and Bulkeley, 2013). Such experiments, as precarious interventions, disrupt socio-technical systems and in doing so call for ongoing maintenance for their continuation. This maintenance is not only material, but also discursive as the narratives that promote and justify low carbon are constantly redefined. The result has potential to create infrastructural configurations that transcend the boundaries and objectives of any low carbon urban experiment, changing the acceptability of low carbon interventions, transforming accepted notions of responsibility, providing scope for formal and informal institutions to engage with low carbon in different ways and opening space for resistance and the alternative appropriation of experiments (Castán Broto and Bulkeley, 2013).

Mechanisms and techniques involved in practising low carbon

The coming together of agents, subjectivities, objects and flows in the making of the low carbon city is orchestrated through a broad spectrum of mechanisms and techniques. These operate as framing and discursive devices – more or less embodied and disembodied – influencing and 'in-forming' the relationships at stake. By mechanisms and techniques we refer to a broad range of technologies of thinking and assembling instruments used in the making of low carbon. These are "devices, tools, techniques, personnel, materials and apparatuses" that enable the possibility to imagine and act upon the conduct of the people to be governed (Miller and Rose, 2008, p. 16; see also Rose, 1999; Murray Li, 2007). Techniques confirm expertise, define an intelligible field of action and roll out specific forms of intervention.

Drawing on the governmentally work of Mitchel Dean (2010), it is possible to argue that, in rolling out specific techniques for advancing a low carbon urbanism, thought takes material from – as charts, texts, graphs, measuring devices, audits, quality controls and other seemingly mundane devices of the everyday are enlisted in the governing effort. Common techniques used in governing the environment are, for example: the production of reports and discourses on the global monitoring of resources (Luke, 1997; Rutherford, 2007); the use of mapping and other representation techniques for the creation of environmental objects (Demeritt, 2001; Braun, 2000; Murray Li, 2007); and the development of systems and indicators to define, classify and separate objects of environmental concern (Bulkeley etal., 2007). In the specific case of the low carbon city, the extensive range of mechanisms and techniques involved in introducing solar energy can be used to illustrate this point. They include local energy baselines and 'solar city' master plans (Indian cities), quality standards for solar hot-water systems (São Paulo), financial instruments developed in the context of the Clean Development Mechanism (Cape Town), green mortgages (Mexico City) and mapping and visualization tools such as solar atlases (Berlin) (Luque-Ayala, 2014; Bulkeley et al., 2014; Luque-Ayala, 2015; Bulkeley et al., 2015).

4 Mobilizing low carbon urbanism

The final dimension of this framework draws on the need to mobilize design and practice; a feat where experimentation, learning and scaling up are often called upon. Urban transitions are always unfolding and in the making. Acknowledging this demands a move away from the notion of successful or failed technological innovations, a centrepiece of first-generation urban low carbon transition studies, to a more processual and socially embedded understanding of the low carbon city. Critical to this conceptual shift is to develop a more nuanced and enlarged understanding of the implications and societal consequences of low carbon urbanism. Here we make reference to various different process involved in mobilizing low carbon urbanism, as a proxy for advancing the question of *how the transition unfolds?* Experimentation, learning and scaling up, by no means exhaustive of the ways in which a low carbon transition takes place, are the result of the constant iteration between designing and practising low carbon, and are seen as steps and moments within an attempt to mainstream a low carbon urbanism. As with other elements within this framework, mobilizing low carbon is not about intervening in a single site or space, but requires changes, transformations and actions across different scales and sites. Each one of the processes discussed in this section connects to modes of development in different manners, and draws differently from their geographical and political contexts.

Experimentation

Increasingly, 'experimentation' is becoming a means through which the governing of climate change is being pursued. To govern urban low carbon transitions remains a key challenge for urban policy-makers, planners and practitioners. In response to the complexities and uncertainties involved, new forms of innovation and experimentation are emerging as a means through which governance can be realized (Bulkeley and Castán Broto, 2013; Frantzeskaki et al., 2014; Truffer and Coenen, 2012). Urban experimentation for low carbon is based on the idea that cities provide a learning arena towards sustainability within which the co-creation of innovation can be collectively pursued among research organizations, public institutions, private sector and community actors (Liedtke et al., 2015; Evans et al., 2015). Experiments "create new forms of political space within the city, as public and private authority blur, and are primarily enacted through forms of technical intervention in infrastructure networks, drawing attention to the importance of such sites in urban politics" (Bulkeley and Castán Broto, 2013, p. 361). Experimentation, when taking the form of demonstration initiatives, tends to be seen as a means for testing technological innovation. Yet, allowing a form of experiential and material learning while enabling various agents to examine the performance and operation of technological artefacts, experimentation plays a role in shifting the forces at play; it operates through and in social processes and governance arrangements; it creates a reverberation, a wave of ripples that affects sites and agents across scales.

Experimentation for urban low carbon transitions seeks to create spaces for innovation and alternatives to be tested and experience gained (Bulkeley and Castán Broto, 2013). Crucially, they provide spaces for innovation and learning beyond purely technological domains: climate innovation and experimentation in cities is as much technical as it is social, political and in forms of governance (Evans and Karvonen, 2014; Bulkeley and Castán Broto, 2013). Experimentation stands as a less directed process than strategic niche management. From 'urban living laboratories' and 'innovation districts' (Voytenko et al., 2016; Evans and Karvonen, 2014) to community and cooperative energy initiatives (Seyfang et al., 2013; Moss et al., 2015), experimentation is seen not only as a means through which to gain experience, demonstrate, and test ideas, but also as a step towards scaling up responses in systems of provision that will have improved effectiveness, political traction and public support. These projects and interventions often bypass traditional funding and planning mechanisms while at the same time create new forms of intervention in the absence of formal policy channels (Hoffmann, 2011).

Climate urban experiments cover a broad range of fronts, from forms of innovation in governance and new modes of social learning to the material transformation of the city's infrastructures. Examples abound, such as exemplar demonstration projects (e.g. the London Hydrogen Bus), iconic 'sustainable' buildings (e.g. Hong Kong's Construction Industry Council Zero Carbon Building in Kowloon), sustainable neighbourhoods and communities (e.g. the Peabody's BedZED housing development in the UK) and 'urban living labs' (e.g. Manchester's Biospheric Foundation). Urban experimentation is not limited to global North cities, as illustrated by Thane, in the Mumbai Metropolitan Region (India), a city that has since 2008 experimented with various techniques aimed at establishing a local governance of energy (Luque-Ayala, 2014); or by Bangalore, where experiments with zero carbon housing test the limits of simultaneously making, maintaining and living low carbon (Bulkeley and Castán Broto, 2014).

Learning and scaling up

The development of shared understanding and learning is critical to how cities respond to climate change. Like experimentation, learning for low carbon takes a multiplicity of forms. It can be about replication, or the transmission of ideas, or it can be about understanding failure. Learning applies to both institutions and individuals, whose exposure to different skills, attitudes, forms of organization and knowledge fosters a form of sustainability learning (Bontenbal, 2009).

A significant amount of scholarly work on urban learning for climate change focuses on the critical role played by transnational municipal networks working on energy, climate and sustainability issues (Kern and Bulkeley, 2009; Marsden and Stead, 2011; Betsill and Bulkeley, 2007). Municipal networks offer a range of opportunities through which learning and knowledge exchange takes place, from the development of best practice case studies to commissioned research, informal peer-to-peer exchange and capacity-building. Many of these opportunities are grounded on the personal exchanges and collaborations fostered by network events. Yet, for municipal staff and other local stakeholders involved, issues of training, employment patterns and other work activities impose limitations on their ability to incorporate the lessons learned in their day-to-day work (Howlett and Joshi-Koop, 2011).

An important aspect of learning urban climate responses concerns the transfer of knowledge between academia and practitioner communities. Within the extensive academic literature on strategic niche management, a range of valuable tools and models for practitioner learning have been developed. By integrating diverse analytical and knowledge domains, these foreground the value of advancing social innovation through managerial perspectives. In this way, they open possibilities for translating "scientific insights into practical valuable guidelines for specific social innovation projects in specific contexts and for specific target groups" (Raven et al., 2010, p. 72; see also Turnheim et al., 2015).

Indeed, translation is also an essential part of scaling up urban climate responses. However, translation is not a single process (Smith, 2007; Seyfang, 2010), but rather a multiple frame through which things become different while still retaining some of their original elements. For example, translation draws directly from the initial (urban) problematization of carbon and climate change: the way in which the climate/carbon 'problem' is framed and positioned within the urban context opens, closes and shapes possibilities for innovation or experimentation (such as the development of niches). Similarly, translation takes shape through a process of integration, where innovation niches gain acceptance through the adoption of regime characteristics. Finally, translation also takes shape via intermediation, where joint projects or partnerships between agents located within both alternative and mainstream spaces (e.g. niche and incumbents) develop new forms (Smith, 2007; Seyfang, 2010). As part of this analysis, the role of innovation niches has been evaluated beyond its nurturing function as a site of empowerment – where the niche is able to either 'fit and conform' into incumbent regimes or 'stretch and transform' them (Smith et al., 2014, pp. 117–118). In thinking about scaling up urban low carbon transitions through translation and empowerment, we can see that "experimentation and socio-technical systems are mutually and relationally constituted," an indication that scaling up requires careful attention to how urban climate responses are "elaborated, sustained and come to be mobilised" (Bulkeley et al., 2015, p. 243). Connecting the three dimensions of this urban low carbon transitions analytical framework, translation and scaling up are revealed as processes constituted through designing, practising and mobilizing low carbon. Translating and scaling up urban low carbon responses means a coming together of ideas (ways of thinking) and forms (material shapes and ways of doing) capable of challenging well-established carbon-intensive ways of going about development and its politics.

5 Conclusion: multiple junctions towards the low carbon city

The coming together of the design, practice and mobilization of low carbon suggests possible pathways through which a low carbon urbanism occurs. Such pathways are neither singular nor linear, but plural, multiple, rhizomatic and interconnected. They should not be seen exclusively as positive, progressive or devoid of conflict. Pathways can also be sites of conflict, contestation or an expression of regressive rationalities. They are multiple junctions where different interests come together.

A critical yet still unanswered question is *how will we recognize an urban transition when we see it?* It has become clear that the transition is not simply a matter of identifying 'best practice' or 'exemplary projects' – both highly political discourses that, in defining objectives, practices and spaces of expertise for the city, align the urban with a particular political economy narrowly focused on global economic competitiveness (Bulkeley, 2006; Hodson and Marvin, 2007). Narratives around 'best practice' generate implementation scenarios where the local appears to be steering the national, while neutralizing opposition, displacing conflicts and relocating tensions downwards (Hodson and Marvin, 2007). Counterbalancing the singular and linear direction advanced by ideas of 'best practice', we argue that a second generation of low carbon urban transition studies needs to be forthcoming in acknowledging multiplicity and difference in the making of sustainability pathways. Agreeing with Rydin et al. (2013), for whom the idea of pathways opens space for a diversity of imagined urban futures to coexist, we see the 'low carbon pathway' as a useful device for understanding the emerging, multiple and often contradictory politics of low carbon systems and its alternative configurations.

This chapter has developed a framework for a second generation of studies of urban low carbon transitions based on a sympathetic critique of the first wave of urban transition studies – as outlined in Table 2.1. The framework has three critical features. First, it shifts away from an emissions-based focus towards a more developmental understanding of the purposes of low carbon urbanism. This calls for a reconceptualization of low carbon in the city as a matter of development modes, where the possibilities for evaluating and advancing urban climate responses are grounded in broader socio-technical configurations and ways of thinking about the low carbon city. Second, rather than focus on transition processes and management, it suggests we need to look more carefully at the agencies and subjectivities, materialities and techniques co-constituted in the practices of low carbon urbanism. Third, it moves away from a narrow focus on successful or failed transitions to develop an enlarged understanding of the implications of experimentation, learning and up scaling. Here, experimentation is not so much a technical endeavour (i.e. testing new technologies) but both a particular form of governing as well as an attempt to create new political spaces. Together, these features seek to foreground the political nature of the low carbon city, while completing a shift away from singular notions of what it means to be low carbon to an acknowledgement of multiple and contested low carbon transition pathways.

The framework is not intended to operate as a rigid and standard template for analyses of low carbon urbanism. Rather, it offers a palette of analytical entry points that, after careful selection and through different combinations, surfaces the complex politics of the low carbon city. All the chapters included in the book have been informed by the overall aim of the framework, and advance this aim through a selection of the issues at stake. For some authors, for example, this has been about prioritizing the ways in which thinking about low carbon drives particular material configurations of the city (e.g. Chapters 3, 4 and 6). For others, it has been about foregrounding the making/role of certain subjectivities in the mobilization of low

TABLE 2.1 Comparing first- and second-generation studies of urban low carbon transitions.

First-generation studies	Low carbon urbanism	Second-generation studies
Extractive model: focus on GHG reduction	Designing	Embedded model of decarbonization: low carbon as a matter of development modes
Transition processes and management	Practising	Agents, objects and techniques
Focus on successful or failed transitions	Mobilizing	Experimentation and learning as the creation of new political spaces; experimentation and socio-technical systems mutually and relationally constituted
Singular pathway to low carbon transition: search for 'best practices'	Overall	Foregrounds the political nature of the low carbon city; acknowledging multiple and contested developmental pathways

carbon urbanism (e.g. Chapters 11, 12 and 13). Overall, all authors have embarked, simultaneously and in different measures, on an analysis of the socio-technical configurations and ways of thinking involved in designing low carbon urbanism, the agents, subjectivities, materialities, flows and techniques involved in practising low carbon, and the experiments and forms of learning associated with mobilizing low carbon. Fostering the possibility of a combination of conceptual entry points in one single analysis, we have avoided structuring the book through the structuring elements of the framework (i.e. designing, practising and mobilizing). Instead, the various chapters of the book are structured in three parts, covering issues of *technologies, materialities, infrastructures* (Part I), issues of *intermediation and governance* (Part II) and *communities and subjectivities* (Part III).

References

Barnett, C., Clarke, N., Cloke, P. and Malpass, A. (2008) The elusive subjects of neo-liberalism. *Cultural Studies*, 22(5), 624–653.

Berkhout, F., Smith, A. and Stirling, A. (2004) Socio-technological regimes and transition contexts. In: Elzen, B., Geels, F. and Green, K. (eds.) *System innovation and the transition to sustainability: theory, evidence and policy*. Cheltenham: Edward Elgar Publishing, pp. 48–75.

Betsill, M. and Bulkeley, H. (2004) Transnational networks and global environmental governance: the cities for climate protection program. *International Studies Quarterly*, 48(2), 471–493.

Betsill, M. and Bulkeley, H. (2006) Cities and the multilevel governance of global climate change. *Global Governance*, 12(2), 141–159.

Betsill, M. and Bulkeley, H. (2007) Looking back and thinking ahead: a decade of cities and climate change research. *Local Environment*, 12(5), 447–456.

Bontenbal, M. (2009) *Cities as partners: the challenge to strengthen urban governance through North-South city partnerships*. Delft: Eburon Uitgeverij.

Bourdieu, P. (1984) *Distinctions*. Cambridge, MA: Harvard University Press.

Braun, B. (2000) Producing vertical territory: geology and governmentality in late Victorian Canada. *Cultural Geographies*, 7(1), 7–46.

Bulkeley, H. (2005) Reconfiguring environmental governance: Towards a politics of scales and networks. *Political Geography*, 24(8), 875–902.

Bulkeley, H. (2006) Urban sustainability: learning from best practice? *Environment and Planning A*, 38(6), 1029–1044.

Bulkeley, H. (2015) Can cities realise their climate potential? Reflections on COP21 Paris and beyond. *Local Environment*, 20(11), 1405–1409.

Bulkeley, H. and Betsill, M. (2003) *Cities and Climate Change: Urban Sustainability and Global Environmental Governance*. London: Routledge.

Bulkeley, H. and Betsill, M. (2005) Rethinking sustainable cities: Multilevel governance and the 'urban' politics of climate change. *Environmental Politics*, 14(1), 42–63.

Bulkeley, H. and Betsill, M. (2013) Revisiting the urban politics of climate change. *Environmental Politics*, 22(1), 136–154.

Bulkeley, H., Carmin, J., Castán Broto, V., Edwards, G.A. and Fuller, S. (2013) Climate justice and global cities: mapping the emerging discourses. *Global Environmental Change*, 23(5), 914–925.

Bulkeley, H. and Castán Broto, V. (2013) Government by experiment? Global cities and the governing of climate change. *Transactions of the Institute of British Geographers*, 38(3), 361–375.

Bulkeley, H. and Castán Broto, V. (2014) Urban experiments and climate change: securing zero carbon development in Bangalore. *Contemporary Social Science*, 9(14), 393–414.

Bulkeley, H., Castán Broto, V. and Maassen, A. (2014) Low-carbon transitions and the reconfiguration of urban infrastructure. *Urban Studies*, 51(7), 1471–1486.

Bulkeley, H., Castán Broto, V. and Edwards, G. (2015) *An Urban Politics of Climate Change*. London: Routledge.

Bulkeley, H., Edwards, G. and Fuller, S. (2014) Contesting climate justice in the city: examining politics and practice in urban climate change experiments. *Global Environmental Change*, 25, 31–40.

Bulkeley, H., Luque-Ayala, A. and Silver, J. (2014) Housing and the (re) configuration of energy provision in Cape Town and São Paulo: making space for a progressive urban climate politics? *Political Geography*, 40, 25–34.

Bulkeley, H., Powells, G. and Bell, S. (2016) Smart grids and the constitution of solar electricity conduct. *Environment and Planning A*, 48(1), 7–23.

Bulkeley, H. and Schroeder, H. (2012) Beyond state/non-state divides: global cities and the governing of climate change. *European Journal of International Relations*, 18(4), 743–766.

Bulkeley, H., Watson, M. and Hudson, R. (2007) Modes of governing municipal waste. *Environment and planning A*, 39(11), 2733–2753.

Callon, M. (1986) Some elements of a sociology of translation: domestication of the scallops and the fishermen of St Brieuc Bay. In: Law, J. (ed.) *Power, Action and Belief: A New Sociology of Knowledge*. London: Routledge & Kegan Paul, pp. 196–230.

Castán Broto, V. and Bulkeley, H. (2013) Maintaining climate change experiments: urban political ecology and the everyday reconfiguration of urban infrastructure. *International Journal of Urban and Regional Research*, 37(6), 1934–1948.

Coenen, L., Benneworth, P. and Truffer, B. (2012) Toward a spatial perspective on sustainability transitions. *Research Policy*, 41(6), 968–979.

Dean, M. (2010) *Governmentality: Power and rule in modern society*. London: Sage.

Demeritt, D. (2001) Scientific forest conservation and the statistical picturing of nature's limits in the Progressive-era United States. *Environment and Planning D: Society and Space*, 19(4), 431–460.

Dowling, R. (2010) Geographies of identity: climate change, governmentality and activism. *Progress in Human Geography*, 34(4), 488–495.

Evans, J., Jones, R., Karvonen, A., Millard, L. and Wendler, J. (2015) Living labs and co-production: university campuses as platforms for sustainability science. *Current Opinion in Environmental Sustainability*, 16, 1–6.

Evans, J. and Karvonen, A. (2014) 'Give me a laboratory and I will lower your carbon footprint!' – urban laboratories and the governance of low-carbon futures. *International Journal of Urban and Regional Research*, 38(2), 413–430.

Feldman, D.L. (2012) The future of environmental networks–governance and civil society in a global context. *Futures*, 44(9), 787–796.

Foucault, M. (2009) *Security, Territory, Population: Lectures at the College de France 1977–1978*. London: Picador.

Frantzeskaki, N., Castán Broto, V., Coenen, L. and Loorbach, D. (2017) Urban sustainability transitions: the dynamics and opportunities of sustainability transitions in cities. In: Frantzeskaki, N., Castán Broto, V., Coenen, L. and Loorbach, D. (eds.) *Urban Sustainability Transitions*. London: Routledge, pp. 1–21.

Frantzeskaki, N., Wittmayer, J. and Loorbach, D. (2014) The role of partnerships in 'realising' urban sustainability in Rotterdam's City Ports area, The Netherlands. *Journal of Cleaner Production*, 65, 406–417.

Fuller, S. (2017) Configuring climate responsibility in the city: carbon footprints and climate justice in Hong Kong. *Area*. doi:10.1111/area.12341.

Gandy, M. (2004) Rethinking urban metabolism: water, space and the modern city. *City*, 8(3), 363–380.

Geels, F. (2010) The role of cities in technological transitions. In: Bulkeley, H., Castán Broto, V., Hodson, M. and Marvin, S. (eds.) *Cities and low carbon transitions*. London: Routledge, pp. 13–27.

Geels, F. (2014) Regime resistance against low-carbon transitions: introducing politics and power into the multi-level perspective. *Theory, Culture & Society*, 31(5), 21–40.

Giglioli, I. and Swyngedouw, E. (2008) Let's drink to the great thirst! Water and the politics of fractured techno-natures in Sicily. *International journal of Urban and Regional Research*, 32(2), 392–414.

Gore, C.D. (2010) The limits and opportunities of networks: municipalities and Canadian climate change policy. *Review of Policy Research*, 27(1), 27–46.

Graham, S. and Marvin, S. (2001) *Splintering Urbanism: Networked Infrastructures, Technological Mobilities and the Urban Condition*. London: Routledge.

Graham, S. and Thrift, N. (2007) Out of order: understanding repair and maintenance. *Theory, Culture & Society*, 24(3), 1–25.

Gram-Hanssen, K. (2011) Understanding change and continuity in residential energy consumption. *Journal of Consumer Culture*, 11(1), 61–78.

Grin, J., Rotmans, J. and Schot, J. (2010) *Transitions to Sustainable Development: New Directions in the Study of Long Term Transformative Change*. London: Routledge.

Hajer, M.A. (1995) *The Politics of Environmental Discourse: Ecological Modernization and the Policy Process*. Oxford: Clarendon Press.

Heinelt, H. and Niederhafner, S. (2008) Cities and organized interest intermediation in the EU multi-level system. *European Urban and Regional Studies*, 15(2), 173–187.

Hodson, M. and Marvin, S. (2007) Understanding the role of the national exemplar in constructing 'strategic glurbanization'. *International journal of Urban and Regional Research*, 31(2), 303–325.

Hodson, M. and Marvin, S. (2009a) Cities mediating technological transitions: understanding visions, intermediation and consequences. *Technology Analysis and Strategic Management*, 21(4), 515–534.

Hodson, M. and Marvin, S. (2009b) 'Urban ecological security': a new urban paradigm? *International journal of Urban and Regional Research*, 33(1), 193–216.

Hodson, M. and Marvin, S. (2010) Can cities shape socio-technical transitions and how would we know if they were? *Research Policy*, 39, 477–485.

Hodson, M. and Marvin, S. (2012) Mediating low-carbon urban transitions? Forms of organization, knowledge and action. *European Planning Studies*, 20(3), 421–439.

Hodson, M., Marvin, S. and Bulkeley, H. (2013) The intermediary organisation of low carbon cities: a comparative analysis of transitions in Greater London and Greater Manchester. *Urban Studies*, 50(7), 1403–1422.

Hoffmann, M.J. (2011) *Climate Governance at the Crossroads: Experimenting with a Global Response After Kyoto*. Oxford: Oxford University Press.

Howlett, M. and Joshi-Koop, S. (2011) Transnational learning, policy analytical capacity, and environmental policy convergence: survey results from Canada. *Global Environmental Change*, 21(1), 85–92.

Hughes, T.P. (1983) *Networks of Power: Electrification in Western Society, 1880–1930*. Baltimore: Johns Hopkins University Press.

Iles, P. and Yolles, M. (2002) Across the great divide: HRD, technology translation, and knowledge migration in bridging the knowledge gap between SMEs and universities. *Human Resource Development International*, 5(1), 23–53.

Kaika, M. (2005) *City of Flows: Modernity, Nature, and the City*. London: Routledge.

Kaika, M. and Swyngedouw, E. (2000) Fetishizing the modern city: the phantasmagoria of urban technological networks. *International journal of Urban and Regional Research*, 24(1), 120–138.

Kemp, R., Schot, J. and Hoogma, R. (1998) Regime shifts to sustainability through processes of niche formation: the approach of strategic niche management. *Technology Analysis & Strategic Management*, 10(2), 175–198.

Kern, K. and Bulkeley, H. (2009) Cities, Europeanization and multi-level governance: governing climate change through transnational municipal networks. *Journal of Common Market Studies*, 47(2), 309–332.

Kooy, M. and Bakker, K. (2008) Technologies of government: constituting subjectivities, spaces, and infrastructures in colonial and contemporary Jakarta. *International Journal of Urban and Regional Research*, 32(2), 375–391.

Latour, B. (2005) *Reassembling the Social: An Introduction to Actor-Network Theory*. Oxford: Clarendon.

Lawhon, M. and Murphy, J.T. (2012) Socio-technical regimes and sustainability transitions Insights from political ecology. *Progress in Human Geography*, 36(3), 354–378.

Legg, S. (2007) *Spaces of Colonialism: Delhi's Urban Governmentalities*. Oxford: Blackwell.

Leigh-Star, S. (1999) The ethnography of infrastructure. *American Behavioral Scientist*, 43(3), 377.

Liedtke, C., Baedeker, C., Hasselkuß, M., Rohn, H. and Grinewitschus, V. (2015) User-integrated innovation in Sustainable LivingLabs: an experimental infrastructure for researching and developing sustainable product service systems. *Journal of Cleaner Production*, 97, 106–116.

Lovell, H., Bulkeley, H. and Owens, S. (2009) Converging agendas? Energy and climate change policies in the UK. *Environment and Planning C: Government & Policy*, 27(1), 90–109.

Luke, T.W. (1997) *Ecocritique: Contesting the Politics of Nature, Economy, and Culture*. Minneapolis: University of Minnesota Press.

Luque-Ayala, A. (2014) Reconfiguring the city in the global South: rationalities, techniques and subjectivities in the local governance of energy. PhD thesis (Geography), Durham University.

Luque-Ayala, A. (2015) Solar hot water and housing systems in São Paulo, Brazil. In: Bulkeley, H., Castán Broto, V. and Edwards, G. (eds.) *An Urban Politics of Climate Change*. London: Routledge, pp. 157–177.

Marino, E. and Ribot, J. (2012) Special issue introduction: adding insult to injury: climate change and the inequities of climate intervention. *Global Environmental Change*, 22(2), 323–328.

Marsden, G. and Stead, D. (2011) Policy transfer and learning in the field of transport: a review of concepts and evidence. *Transport Policy*, 18(3), 492–500.

May, T. and Marvin, S. (2009) *The SURF/ARUP Framework for Urban Infrastructural Development: A Joint Report Between SURF and Ove ARUP for the ESRC Business Placement Fellows Scheme*. Salford: SURF – Centre for Sustainable Urban and Regional Futures.

McFarlane, C. and Rutherford, J. (2008) Political infrastructures: governing and experiencing the fabric of the city. *International journal of Urban and Regional Research*, 32(2), 363–374.

McGuirk, P., Bulkeley, H. and Dowling, R. (2014) Practices, programs and projects of urban carbon governance: perspectives from the Australian city. *Geoforum*, 52, 137–147.

Meadowcroft, J. (2007) Who is in charge here? Governance for sustainable development in a complex world. *Journal of Environmental Policy & Planning*, 9(3–4), 299–314.

Miller, P. and Rose, N. (2008) *Governing the Present*. Cambridge: Polity.

Monstadt, J. (2009) Conceptualizing the political ecology of urban infrastructures: insights from technology and urban studies. *Environment and planning A*, 41(8), 1924–1942.

Moss, T. (2011) Intermediaries and the governance of urban infrastructures in transition. In: Guy, S., Marvin, S., Medd, W. and Moss, T. (eds.) *Shaping Urban Infrastructures: Intermediaries and the Governance of Socio-technical Networks*. London: Earthscan, pp. 17–34.

Moss, T., Becker, S. and Naumann, M. (2015) Whose energy transition is it, anyway? Organisation and ownership of the Energiewende in villages, cities and regions. *Local Environment*, 20(12), 1547–1563.

Murray Li, T. (2007) *The Will to Improve: Governmentality, Development, and the Practice of Politics*. Durham, NC: Duke University Press.

Raven, R., Kern, F., Smith, A., Jacobsson, S. and Verhees, B. (2016) The politics of innovation spaces for low-carbon energy: introduction to the special issue. *Environmental Innovation and Societal Transitions*, 18, 101–110.

Raven, R., van den Bosch, S. and Weterings, R. (2010) Transitions and strategic niche management: towards a competence kit for practitioners. *International Journal of Technology Management*, 51(1), 57–74.

Rice, J.L. (2010) Climate, carbon, and territory: greenhouse gas mitigation in Seattle, Washington. *Annals of the Association of American Geographers*, 100(4), 929–937.

Rose, N. (1999) *Powers of Freedom: Reframing Political Thought*. Cambridge: Cambridge University Press.

Rutherford, S. (2007) Green governmentality: insights and opportunities in the study of nature's rule. *Progress in Human Geography*, 31(3), 291–307.

Rutland, T. and Aylett, A. (2008) The work of policy: actor networks, governmentality, and local action on climate change in Portland, Oregon. *Environment and Planning D: Society and Space*, 26(4), 627–646.

Rydin, Y., Turcu, C., Guy, S. and Austin, P. (2013) Mapping the coevolution of urban energy systems: pathways of change. *Environment and planning A*, 45(3), 634–649.

Schot, J. and Geels, F.W. (2008) Strategic niche management and sustainable innovation journeys: theory, findings, research agenda, and policy. *Technology Analysis & Strategic Management*, 20(5), 537–554.

Seyfang, G. (2010) Community action for sustainable housing: building a low-carbon future. *Energy Policy*, 38(12), 7624–7633.

Seyfang, G., Park, J.J. and Smith, A. (2013) A thousand flowers blooming? An examination of community energy in the UK. *Energy Policy*, 61, 977–989.

Shove, E. and Walker, G. (2007) Caution! Transitions ahead: politics, practice and transition management. *Environment and Planning A*, 39(4), 763–770.

Smith, A. (2007) Translating sustainabilities between green niches and socio-technical regimes. *Technology Analysis & Strategic Management*, 19(4), 427–450.

Smith, A., Kern, F., Raven, R. and Verhees, B. (2014) Spaces for sustainable innovation: solar photovoltaic electricity in the UK. *Technological Forecasting and Social Change*, 81, 115–130.

Smith, A. and Raven, R. (2012) What is protective space? Reconsidering niches in transitions to sustainability. *Research Policy*, 41(6), 1025–1036.

Smith, A., Stirling, A. and Berkhout, F. (2005) The governance of sustainable socio-technical transitions. *Research Policy*, 34(10), 1491–1510.

Smith, A., Vo , J.P. and Grin, J. (2010) Innovation studies and sustainability transitions: the allure of the multi-level perspective and its challenges. *Research Policy*, 39(4), 435–448.

Strengers, Y. (2013) *Smart Energy Technologies in Everyday Life: Smart Utopia?* London: Springer.

Swyngedouw, E. (2006) Circulations and metabolisms: (hybrid) natures and (cyborg) cities. *Science as Culture*, 15(2), 105–121.

Swyngedouw, E. and Kaika, M. (2000) The environment of the city … or the urbanization of nature. In: Bridge, G. and Watson, S. (eds.) *A Companion to the City*. Oxford: Blackwell, pp. 567–580.

Truffer, B. and Coenen, L. (2012) Environmental innovation and sustainability transitions in regional studies. *Regional Studies*, 46(1), 1–21.

Turnheim, B., Berkhout, F., Geels, F., Hof, A., McMeekin, A., Nykvist, B. and van Vuuren, D. (2015) Evaluating sustainability transitions pathways: bridging analytical approaches to address governance challenges. *Global Environmental Change*, 35, 239–253.

UN Population Division (2014) *World Urbanization Prospects – The 2014 Revision*. New York: United Nations, Department of Economic and Social Affairs.

van Lente, H., Hekkert, M., Smits, R. and van Waveren, B. (2003) Roles of systemic intermediaries in transition processes. *International Journal of Innovation Management*, 7(3), 247–279.

Voytenko, Y., McCormick, K., Evans, J. and Schliwa, G. (2016) Urban living labs for sustainability and low carbon cities in Europe: towards a research agenda. *Journal of Cleaner Production*, 123, 45–54.

While, A. (2014) Carbon regulation and low carbon urban restructuring. In: Hodson, M. and Marvin, S. (eds.) *After Sustainable Cities*. London: Routledge, pp. 41–58.

While, A., Jonas, A. and Gibbs, D. (2010) From sustainable development to carbon control: eco-state restructuring and the politics of urban and regional development. *Transactions of the British Institute of Geographers*, 35(1), 76–93.

PART I

Technologies, materialities, infrastructures

3

SEEKING EFFECTIVE INFRASTRUCTURES OF DECARBONIZATION IN PARIS

Material politics of socio-technical change

Jonathan Rutherford

1 Introduction

This chapter is an attempt to make sense of the nature, modalities, outcomes and possibilities of urban transition in Paris. This is a low carbon transition in the sense that there is much work going on in and around the city on different levels to address climate and energy issues and to meet various decarbonization goals. But this climate and energy work always intersects with, and indeed trades off against, a variety of other equally meaningful concerns and strategic objectives. There is then a shifting interplay in Paris between agents and agencies, objects and flows, and mechanisms and techniques that may be consistent with 'practising low carbon', and which it is important to explore in order to develop an understanding of "what and who is involved in the transition" (Luque-Ayala et al., Chapter 2, this volume). This would seem to necessitate a double, intertwined move.

First, it is very tempting to read what is going on in Paris on the low carbon front in relation to, or in comparison with, other cities where decarbonization (in a narrow sense) is more advanced or has achieved more tangible results. This would lead to a focus on the factors or reasons for the relative lack of success or for the belittled place of Paris on some generic global low carbon trajectory or pathway (the inertias, the path dependencies, the obstacles, the obduracies ...). Figures suggest that CO_2 emissions in Paris are indeed stable rather than declining (see Mairie de Paris, 2012a), and that, whether it is the 20% reduction by 2020 on a European level, the Factor 4 reduction by 2050 on a national level, and/or the 25% reduction by 2025 on a municipal level, Paris is therefore not decarbonizing (fast) enough. The chapter eschews this view of Paris as untransitioning or not transitioning, which is based on a narrow, homogeneous, univocal ('end-of-pipe') definition or understanding of low carbon or decarbonization in which a singular quantitative end point, result or desire (drastically reduced CO_2 emissions) can be used to capture or analyse the significance of the far wider, more complex, and sometimes hidden set of processes and practices through and for which 'low carbon' is actually being done (cf. Luque-Ayala et al., Chapter 2, this volume). There is a fundamental tension or controversy between different actors, groups and interests in Paris over even defining and framing the issue, as well as over the content and processes of operation of any 'transition'.[1] In this context, we need to avoid foreclosing debate around what might constitute a productive site, arena or practice of future urban change. A more processual perspective is adopted here to privilege exploration and excavation of where, how, why and by/for whom transition work is (actually or potentially) being done, thus unsettling and disrupting habitual understandings and frames of reference focused solely on extracting carbon from just the official record of Paris emissions.

Second, this focus on ongoing processes and practices around low carbon, irrespective of measurable results, means thinking through, even perhaps rethinking, the relationship between the 'what' and the 'who' in transition – specifically, the role of urban materiality in socio-technical change. It means focusing on overlaps, blurrings and recombinations of actors, objects and technical configurations in a co-productive process or a space of coalition or alliance-building wherein it is difficult to disentangle or to define any particular individual component as just an actor, an object or a configuration without referring to their relational make-up and effects. Practising low carbon always involves modes of "ordering, circulating and manipulating" things (Latham et al., 2009, p. 57), i.e. ways of acting on and reconfiguring urban materiality for particular interests, which is an inherently political process, so that struggles over what matters or making things matter have to be seen in a conjoined material and political sense. Low carbon transition comes to matter through the constitution, mobilization and enrolment of things, objects and flows for particular purposes and interests which are (the things, the enrolment, the purposes) inherently contested (by others). Understanding how this enrolment is done (how things are put to work and therefore constituted as matter) and how it becomes contested says a great deal not just about the active role of matter – or stuff – in systemic change, but about the nature and possibilities of materiality as multiple, inherently disputed forms of relationship between us and our physical world. This is, after all, what energy and carbon issues and debates are fundamentally about.

2 Materiality, politics, transition

A lot of work has recognized and highlighted the multiple, contested roles of urban infrastructures in operating energy and low carbon transitions (see for example many of the chapters in Bulkeley et al., 2011). This work focuses on the evolutions and reconfigurations of infrastructures (thereby seeing the latter as dynamic rather than fixed and static), and sees the materialities of infrastructures as emerging as much in the socio-political negotiations they demand as in the actual technical deployment of physical networks. Understanding infrastructure thus always involves taking into account a wider set of materials than the basic, physical, apparently inert equipment that makes up the networked systems underpinning the urban environment. This set of materials cannot just be constituted by following a broader variety of objects and things as they are mobilized in policy-making and contestation. It invokes more a need to open out notions of urban materiality to account for the multiple settings and arenas in flux through and in which urban energy–climate policies and issues are materialized and transformed – "taking account of the distinctive kinds of effectivity that material objects and processes exert as a consequence of the positions they occupy within specifically configured networks of relations that always include human and non-human actors" (Joyce and Bennett, 2010, p. 5). A focus on urban materiality is a focus directly on the contested processes and practices of change, because the diverse ways in which people understand and engage with the shifting sites and arenas of negotiation constantly work and rework the material constructions and experiences of their living space.

Matter and materiality cannot be taken at face value as objects or stuff with pre-existing, stable and shared meanings, significance and implications, but explored instead as a political process "in which the more tangible, physical stuff of the city is a lively participant" (Latham et al., 2009, p. 62). This means that the nature and contours of transition are always ongoing and continuously constituted (pushed together, pulled apart) by the ways in which particular urban artefacts and bits of infrastructure become of interest to actors and groups and become disputed as they are put to work for divergent interests and purposes (Rutherford, 2014). Prescribed policy goals, visions and actions can therefore be reinterpreted (or misinterpreted), reiterated and contested by the different groups and interests present according to what matters to them. Urban materiality thus becomes a key arena for urban politics as a set of "everyday struggles over ecological (re)production and consumption" (MacLeod and Jones, 2011, p. 2450).

In this way, introducing some of the recent thinking around urban materiality into debates around energy and climate issues complements and extends existing work around strategic infrastructure and systemic socio-technical change by showing how the process of making things matter is an inherently contested operation or encounter of multiple coexisting engagements with the concrete objects, natures and flows of the urban living space. What this approach to materiality offers for the study of urban energy–climate issues is a more precise understanding of the disparate settings and arenas in and through which policy discourse and goals are actively translated into actual concrete actions and political interventions. These material settings and arenas are constituted by shifting relations and engagements between multiple urban actors and all kinds of objects and infrastructures. They are also therefore inherently constitutive of (the potential for) socio-political struggles as the orientations of ongoing and future urban ecological transitions are materially understood, experienced and performed in diverging ways.

From this perspective, the French energy system and energy–climate policy context is not a "passive and stable foundation on which [urban] politics takes place" (Barry, 2013, p. 1). Rather, low carbon politics in Paris are actively produced and indeed constituted by artefacts and relations of this system, which become fundamental to the conduct and possibilities of politics. Many of the tensions and struggles over Paris's socio-technical systems emerge around the flows, infrastructures and materialities which make up this system and its functioning. There is neither a pregiven shared understanding of the nature of the system, and its 'basic' components and interactions (for example, what the Paris district heating network or electricity system is), nor agreement over dispositions and configurations of the system for 'sustainable' present and future provision of the French capital. There is, instead, a series of competing accounts (information) of the nature, meaning, significance and future of heat, electricity and smart technology for Paris. This constitutes a 'political situation' around low carbon transition, with indeterminate and contingent bounds and significance, which calls for "an expansion of the range of elements that should be considered when analyzing a controversy ..." (Barry, 2013, p. 11). So the aim here is in part to follow Barry in "developing accounts of the political geography of materials whose ongoing existence is associated with the production of information" (Barry, 2013, p. 5). Then we might see in all this information (about logistics/circulation of materials and flows of power) not consensus and shared visions of low carbon futures but "new concerns, sites and problems about which it matters to disagree" (Barry, 2013, p. 5), and therefore new political possibilities forged from engagement with carbon materialities.

3 Negotiating transition matter(s): active infrastructures and contested change

As the head of the Paris climate agency has suggested, the work of local policy actors in and around the city can only influence, at best, about 25% of its emissions (Ged, 2015). This means looking elsewhere for more distributed sites of, and possibilities to effect, change.

In this section, I focus on three infrastructural–urban material arenas wherein part of the salience of 'the political situation' of low carbon transition is negotiated: in heating systems, in electricity distribution and in smart technology deployment. These reflect a focus in Paris on the technical and political enactment of transition through debating/making changes to socio-technical systems. Other arenas are available – energy efficiency and building retrofit, eco-urban projects, air quality, transport systems and mobility, etc. – and, by exploring these instead, some of the objects and flows under study would change, but there would still be a political process of materially constituting transition possibilities, which is the basis of what I'm trying to get at here.

Heat frictions: pipes, resources and 'Gruyère politics'

Paris has one of the oldest and most extensive district heating systems in Europe. There is an explicit policy objective both on a city and on a regional level to enlarge and interconnect the district heating system to increase economies of scale and to diversify and improve the flexibility of the system (e.g. render it more resilient by using more than one resource). Paris municipal plans and actors make reference to using its position as owner of the heat infrastructure and co-owner of the CPCU distribution company (Compagnie Parisienne de Chauffage Urbain) to extend the network to other parts of the city currently heated by less efficient fuel or electricity (even if the latter is better from a carbon viewpoint) (Mairie de Paris, 2012b). The regional-level climate air and energy master plan (SRCAE) of December 2012 outlined an ambitious goal for a 40% increase in connections to heating systems by 2020, representing 450,000 extra buildings, to be achieved by extension of existing systems and interconnection of networks (CRC d'Île-de-France and PR d'Île-de-France, 2012). But how these objectives are to be met, and how the current heating system can be reconfigured for transition, are contentious issues.

It is not clear for many actors how the goal of 40% more connections to heating systems by 2020 is to be achieved. As of 2015, growth in connections was around 1.5% per year, which is a long way from the goal, such that a real connection policy needs to be outlined and implemented (heating company official, interview). It is not evident, either, which is the best level for investment and heat load. Many suggest a collective and mutualized approach, albeit across different territories (Paris and the inner ring of municipalities, or the new Paris Métropole inter-municipal cooperation structure …), but this would require somebody to take a lead and to organize things, and urban governance has been in a state of flux in Paris in the 2010s (various interviews).

Furthermore, heating systems are heterogeneous and contentious bits of urban materiality. Objectives of interconnection, mutualization and economies of scale require overcoming the technical and contractual difficulties of linking together different systems. The CPCU Paris-centred network is a steam system, while systems in surrounding municipalities use hot water. There are thermodynamic limits to the extent to which energy can be transferred between a hot-water system at 80°C and a high-pressure steam/vapour system at 230°C. These different technologies were one reason (along with local political tensions) preventing the Batignolles planning project in the north-west of the city sending heat "easily and cheaply" across the municipal border into neighbouring Clichy Levallois (City of Paris official, interview). Interconnection of heating systems is also problematized by the different types of existing contracts through which municipalities have conceded their own systems to particular operators, e.g. distribution only as in the case of Paris or also production as in other municipal contracts. The lengths of each contract and the dates when they are up for renewal also vary greatly. French public contract reg-ulations (the *Code des marchés publics*) prevent municipalities from modifying or stopping a particular contract, so combining or mutualizing separate systems across municipal boundaries would require, in theory, separate contracts to be up for renewal at the same time and to have similar conditions of operation. Interviewees referred to the 'tangle' or 'minefield' (*maquis*) of contracts stopping what would appear to be common-sense technical connections such as the use of recuperated heat from EDF's (Électricité de France's) Ivry electricity plant to the east of the city, which is just lost (City of Paris works department official, interview).

Heating pipes are also presented as a political nightmare. They are placed under the road rather than the pavement (due to their diameter) and so necessitate costly major street works for their repair and maintenance. These works also require permits from local authorities who are rarely keen for streets to be regularly dug up due to the unpopularity with local residents. The City of Paris works department cites this local opposition as an important factor on the extension and maintenance of the CPCU system, which can lead in some areas where there has been an accumulation of works to local politicians blocking any new street works (City of Paris works department official, interview). Digging up roads, even for urgent maintenance, is "almost impossible" for a twelve-month period prior to local elections (heating company official, interview). A major biomass plant project in the north-east of Paris, which was a major part of plans for further decarbonizing Paris's heating system, was abandoned when it was estimated that laying the pipes linking the plant to the existing CPCU network would have been as costly as building the plant itself due to their traversing five municipalities. For at least one interviewee using the analogy of a hard cheese full of holes, this inherent degree of permanent contestation represents the substantial difficulties of working with the Paris "Gruyère" (heating company official, interview).

Another area of uncertainty and tension in heating systems concerns ways to drastically increase the proportion of renewable and recuperated heat in the energy mix of systems. It is not clear, for many actors, how this can be achieved. In spite of the use of just over 40% waste incineration (recuperated energy) in CPCU's energy mix, the company has found it difficult to reach the level of 50% renewable and recuperated heat. This threshold triggers both a decrease in VAT level from 19.6% to 5.5% (thus permitting a reduction in customer tariffs), and a process of so-called 'classification' of networks allowing the City of Paris to force new buildings and planning projects to connect to existing heat networks which is an important element in ensuring a return on infrastructure investments.

Biomass resources have represented around 3% of the mix in the region (or 0.1 Mtep (million-ton equivalents of petroleum)), yet this is supposed to increase to 30% of the mix (or 1.2 Mtep) by 2020 according to the SRCAE. This increased use of biomass is a primary instrument to meet local climate and energy plan objectives. Plans are to use an existing plant and replace half the coal with wood pellets that are not humid and burn quite easily, and can be used in the same boilers. But this demands an increase in logistical capacity for transport and storage necessitating substantial investment given current limits to local availability and access to biomass resources. The wood will therefore come initially at least from Canada and Ukraine because they have an established supply chain and industry which France does not yet have. Some observers unsurprisingly question the carbon efficiency of long distance imports of wood to meet climate and energy goals (various interviews). Furthermore, in the longer term, the Paris region is not the only French region orienting its climate and energy plan around biomass, leading to prospective competition for resources where France is unlikely to be able to produce enough wood for all (City of Paris works department official, interview). Another regulatory constraint is the fact that some biomass components such as wood from construction projects and wood pallets are classed as 'waste' by French legislation, and so are subject to a restriction on their treatment. Some actors denounce the contradictions of a system which is supposed to be trying to create a biomass industry (senior adviser to City of Paris politician, interview), and yet is exporting perfectly usable wood pallets to Sweden.

In the SRCAE, geothermal is also supposed to increase from 3% to 13% of the regional energy mix by 2020. But geothermal energy requires a lower temperature heat network than CPCU's vapour network (City of Paris works department official, interview). There can also only be one geothermal well or sink within a certain area to avoid over-extraction problems which can create inter-municipal problems if the area crosses boundaries (City of Paris official, interview) or even competition over what is actually a limited resource, even though Île-de-France sits on the large Dogger aquifer (City of Paris works department official, interview). The heavy initial investment required for either large-scale biomass or geothermal production appears to be a significant barrier (heating company official, interview). Networks using these resources would need to connect to thousands of homes to guarantee any kind of heat load and a return on investment, and the lack of this demand base has led to abandoning of possible projects (City of Paris official, interview).

In sum, resource flows, infrastructure adjustments, contractual and regulatory issues and their techno-political interpretation and mobilization constitute major arenas of dispute over how to organize provision of low carbon heat for Paris to meet its climate and energy objectives.

Electricity distribution as public knowledge controversy

After many years of being taken care of behind the scenes, electricity distribution has become a local matter again in Paris and other French cities. The City of Paris owns its electricity distribution system but it has been run by EDF since 1955 under the terms of a 55-year public service concession contract. Following electricity sector reforms, vertical separation of activities led to the creation of a wholly owned subsidiary ERDF (Électricité Réseau Distribution France) in 2008 for distribution in most of France (as a local public service monopoly) with EDF concentrating on production and sale.[2] The distribution contract came up for renewal at the end of 2009 in a context of debate over apparent lack of EDF/ERDF investment in the city's electric grid, which was linked by the press to a series of blackouts and power cuts in the city (Bezat, 2010). The contract became the core element in technical, administrative, financial and legal debate and wrangling over urban electricity provision. It was notably a key point of focus for local politicians, tasked with implementing a local climate–energy plan, to retake an interest in and regain some control over a system and a public service that had become shaped largely to meet the particular interests of EDF. High tensions emerged over a low tension grid, as shown by the detailed investigations and reports produced by national and regional public accounting courts,[3] the complex legal and administrative procedure, and the mediatization of the stakes, process and outcome.

The process of negotiation over the renewal of the contract became a dispute over the actual system itself both in terms of ownership of some of its components, and control of the system and the interests which were to be privileged through this control. In particular, there were diverging views of whether this was Paris's network (to be shaped according to the needs of the capital) or a local network within ERDF's national system (with the maintenance and safeguarding of the latter as the primordial concern).

The City of Paris remains the owner of all infrastructure and equipment between the distribution substations, where voltage is stepped down from RTE's (Réseau de Transport d'Électricité's) transmission grid, up to and including customer meters. However, it emerged that there were some divergences over bits of property and infrastructure as an inventory had not been kept up to date. This is a crucial question for accounting, for legal issues and for calculation both of investments required in the delegated system and the annual return or royalty (*redevance*) that the municipality obtains based on the performance of ERDF, and which Paris redistributes to help poorer households pay their bills. There was dispute over the financial value and provisions for investment and renewal of the ageing, amortized infrastructure, and the complex accounting methods for calculating these. The Regional Court of Accounts criticized the mode of calculation used by ERDF (CRC d'Île-de-France, 2010, p. 3). This was linked to contestation over ownership of some assets (meters, substations ...) (Baupin and Gassin, 2009), and their value or rate of depreciation, which directly affects investments which are calculated according to the gross value of assets (FNCCR, 2015, p. 39). The investment value of the assets (and thus the provision made for maintenance and renewal of the Paris system) in EDF/ERDF's accounting rapidly declined from around 1 billion euros (EUR) at the end of 2001 to only 350 million EUR at the end of 2009 (i.e. the investment required to return the infrastructure to the City in its 'original' state would be three times less). Local politicians wrote brutally, and appeared on DailyMotion (see Énergie2007.fr 2015), concerning 'pillaging', 'hemorrhage' and 'evaporation' of money from the Paris system into the accounting books and results of EDF on a national level (Baupin and Gassin, 2009).[4]

As well as a dispute around control over electricity distribution and ownership of parts of the system, this was at the same time a dispute about knowledge of the system, transparency and circulation of information. The City refused to accept and validate EDF's annual reports (for both public services of distribution and supply) because they did not contain sufficient technical and financial information to allow it to monitor the evolution of its system. When ERDF carries out investment and maintenance work in the distribution system, the City is totally reliant on ERDF for information about the nature and extent of the work. Yet, the Regional Court of Accounts found that annual reports "contained little information about the means, technical and human notably, put in place by EDF to reach its objectives" (CRC d'Île-de-France, 2010, p. 13). ERDF is working on a national level with hundreds of concessions and the technical functioning and interconnection of distribution networks means it is difficult to extract and provide information about one local system (Cour des Comptes, 2013, p. 121). There is a knowledge controversy then about the scale (*maille*) of extraction and provision of information, or over where the boundary might be between a Paris 'inside' and a national 'outside'.[5] So before taking a decision on the prolongation of the contract at the end of 2009, and the prospective terms of this contract, the City conducted three 'internal' audits, itself, of the state of affairs for technical, property and financial-legal concerns. As the City of Paris politicians involved in negotiations observed understatedly at the time: "After decades of unilateral management and within a framework which doesn't give much room to local authorities, reaffirming the rights of the City has not been without trouble" (Baupin and Gassin, 2009, p. 2).

Yet, notably given the uncertainty over the continuing regulatory status of distribution monopolies, the City of Paris was unprepared to do anything other than adjust the terms of its contract with EDF/ERDF. A compromise was eventually negotiated leading to a prolongation of the existing contract for fifteen years (itself contested by a competing electricity company), albeit with a number of clauses clarifying investments and procedures, regarding provision of information notably. It also opened the possibility for the City to potentially remunicipalize the system and service before then if regulations and conditions were to allow, which is viewed as important in safeguarding local capacity in the context of the need to constantly track and update evolving Paris climate–energy plans in the coming years (Ville de Paris et al., 2009).

This conflict involving contract negotiation, accounting reports, legal wrangling and technical expertise brought an amortized infrastructure back to life, and recreated a degree of local capacity and control over a socio-technical system in low carbon transition.

Peak loads, smart fictions and 'associating heterogeneous actors'

Paris requires ever more 'intelligent' work to tie together or conjoin production and consumption of electricity. This has, to some degree, long been the case (see Beltran, 2002), but it was also recognized in the negotiations over extending ERDF's distribution contract (Ville de Paris et al., 2009, pp. 10–12). Île-de-France now imports more than 90% of its electricity, demand for power is projected to increase, and the peak load charge (*appel de puissance*) is already increasing much more rapidly than actual consumption. This means that during particular short periods (in winter) there is a punctual boost in demand which places stress on the electricity system to deliver the required amount of power. Renewable power is often small scale and intermittent in its supply, and at peak times, the French electricity system relies on carbon-intensive production which goes against climate goals and does little for reducing regional energy dependency. The system must therefore contains the capacity to generate, transport and distribute sufficient supply to meet these occasional peaks, as well as the knowledge and expertise to be able to monitor and anticipate as closely as possible when and where these are likely to take place. This has become a crucial issue for energy in the Paris region.

A number of overlapping measures and actions have been initiated as a response, with an overarching aim of flattening the load charge (*lissage de la courbe de charge*) and "better consumption of electricity" (ERDF official, quoted in ERDF, 2011, p. 52). Real-time demand-side management measures encourage, promote or oblige, for example, particular blocks of consumption to be reduced or, at least, displaced in space and/or time. Some tasks might be undertaken during the night for example when demand is low, e.g. "use the inertia of a building to heat it outside of peak times" (APUR, 2013, p. 9) thus reworking the relation between urban materiality/inertia, time and network.

Smart meters are being rolled out to provide more data and information for ERDF, RTE and EDF about consumption patterns and quantities. Smart grid projects are beginning to be proposed to experiment with mutualization of demand – "agglomerating several consumption models" (ERDF official, quoted in ERDF, 2011, p. 52) – and of local production. It is held that buildings and neighbourhoods can be interconnected to share local renewable energy or to group loads together to propose more coherent, and less heavily differentiated, temporal electricity demand.

The IssyGrid project to the south-west of Paris is a prime example of this 'experimental' approach to 'optimizing energy consumption' to "consume better, less and at the right time" (Ville d'Issy-les-Moulineaux, n.d.). This takes the form of making 'real-time' adjustments between energy production at renewable energy installations in two city districts and increasingly variable energy demand, offering a local response to a wider need for flexible peak load management. Reflecting its motto where "to govern is to foresee," the Issy municipality (and its ten big company collaborators and associated start-ups) are increasing the controllability and predictability of local energy flows by deploying integrated systems for the production of information. As well as speaking to an 'autonomous city' discourse, various distinct logics coalesce around the IssyGrid project creating a socio-technical coalition of political-economic interests brought together by 'smart': a playground for testing/experimenting new technologies, devices, techniques and business models (corporate interests), load management (EDF, ERDF), reduced energy bills (Issy municipality), city branding and publicity with internal (political) and external (growth) objectives (Issy municipality).

The result is an electricity system configuration supported by detailed real-time information circulation, temporal adaptability both of supply and demand (both during the day and between seasons), new "associations of heterogeneous actors" (ERDF official, quoted in ERDF, 2011, p. 52), insertion of new technologies and meters for 'optimal' performance of the system (as a shared goal irrespective of particular interests), and closing distances between domestic electricity use and practices and power production elsewhere. This is a new socio-technical complex which reworks space, time and technology ostensibly for user-oriented goals but may be seen as a configuration to sustain (growth and expansion of) an existing system. As the IssyGrid project stakeholders argue: "This realistic approach allows us to deploy at least cost a territorial energy optimization policy without questioning previous or future infrastructure choices" (IssyGrid Project, n.d.).

4 Effective materiality

These three brief vignettes highlight something of the use of debates over and practices of infrastructure reconfiguration in decarbonization work in Paris. They also illustrate the inherently contested nature of the ongoing making of low carbon urban fabrics. More technology, more production and circulation of information and data, and more knowledge and reflexive feedback in socio-technical systems does not reduce the possibility or the intensity of dispute and disagreement (it does not produce consensus), but instead leads to new sites and matters of politics with indeterminate (and contingent) bounds and significance (Barry, 2013, p. 10).

The three arenas constitute particular knowledge controversies about or around infrastructure, or particular bits/materials of infrastructure, through which a wider 'political situation' around the nature, modalities, outcomes and possibilities of low carbon transition are being debated, negotiated, effected and contested. Disputes emerge and circulate around pipes, resources and energy mixes, contracts, finance, meters and urban projects, recombining bits of infrastructure and the urban fabric with the production of information and circulation of knowledge about these, as well as with stories, narratives and discourses about urban futures which together become highly politicized. They emerge in or through different spaces which reflect at once the technical functioning of infrastructures, the presence of distinctive forms and sources of knowledge and expertise about these, and often diverging techniques of communication or circulation of opinion and information (in private negotiations between stakeholders, in legislative arenas, on DailyMotion and in the press, in public presentations, etc.).

The political constitution and operation of a low carbon transition thus cannot be dissociated from, and indeed works through, a host of otherwise rather prosaic material encounters and connections, such as the interlinking of steam and hot water heating systems with different thermodynamic properties, the accounting techniques for valuing an electricity distribution network and how this impacts future reinvestment ratios, and the deployment of a real-time information system integrating data on energy production and consumption. Each material encounter filters, mediates between and translates varying interests, visions and actions, thus contributing effectively to shaping the realm of what is possible in doing transitions. The contours of local knowledge controversies – about the technical functioning, economic calculations, modes of transparency, circulation of information, etc. – make up the meaning of the political situation.

The heating, electricity and smart arenas examined also suggest differing ways in which, or processes through which, infrastructure comes to matter or to be opened to controversy, i.e. when their significance, value, operation or behaviour becomes of interest to particular groups. Three examples of the distinctive effectivity of infrastructure can be sketched here.

First, there is something here about *demonstration* and transparency, about highlighting certain things (and not others), and about where the boundary lies between visible and invisible, public and private, or something that matters and something to be ignored. Information has come to light in all three arenas and helps to open debate on certain points, but other information is withheld such that the realm of the politics of infrastructure is inherently bounded by accessibility of data. Public demonstration in the case of the IssyGrid project or the information circulated by Paris politicians about the electricity distribution contract places matters in the public record with a view to producing agreement and accord, but these become subject to debate. Barry, examining the material politics of infrastructures, showed that the more information circulates and the more issues are discussed and debated, the more disagreement and dissensus is produced, new sites and arenas of politics result … and indeed "could virally multiply" (Barry 2013, p. 182).

Second, in a process of *revendication*, materials become open to airings of claims or possibilities as to what (or indeed whose) they are or what they stand for. Objects are seized or appropriated as part of a territorial strategy or an aim to make things work for specific purposes, and in some cases they come to exceed their immediate constitution. In the three arenas, this is especially the case at particular times (when contracts are up for renewal, in accordance with changing legislation and regulations) or through particular calculations (for lowering VAT, market prices of coal and gas, or intermittence of renewable energy production vs grid load charge). These may be seen to represent shifting and contested 'adequations' (spatial, technological, institutional-territorial) as actors seek new commensurations between production and consumption, control and concession, density and sprawl, the present and the future, etc.

Third, actors must take seriously the active role and *recalcitrance* of matter and materials in rendering infrastructure configurations open to debate. Components or elements of heat, electric and smart systems have certain qualities or behave in certain unruly ways such that there is some difficulty in mastering or making sense of different matters, objects or understandings or views of objects and their performances and operations. Paying more attention for example to the intrinsic thermodynamic properties of energy, and their political consequences, becomes essential (Barry, 2015). Accepting this recalcitrance of matter implies that some degree of agency is devolved to obstreperous objects, and that actors must adjust their strategies as a consequence.

The politics of urban energy and climate change emerge not just over infrastructure and concrete objects per se, but more specifically in the processes of overflowing of these infrastructures and objects, and therefore the ways in which urban materiality is inherently manipulated through the practices and performances of varying groups and interests. Taking into account the multiple flows (of emissions, heat, people, money, ideas …) related to urban energy–climate issues and the different orderings and disorderings through which they circulate helps to disrupt the linear pathways which normative transition discourse proclaims and enacts. Unpacking the diverse and undulating processes through which energy and climate issues effectively come to matter in the urban arena is thus a useful means of tracing how transition is being performed, contested and repoliticized.

5 Conclusion

> To change the world, you have to "interlace it with steam and electricity!"
> (Mattelart, 1999, p. 179, quoting the nineteenth-century French engineer Michel Chevalier)

Low carbon stakeholders in the city have become differently organized around, and indeed by, particular material sites, objects, resources and infrastructures in attempts at creating and sustaining visions and implementing actions towards urban transition. This chapter has explored some of the diverse array of material artefacts, sites and arenas around which low carbon actions and developments in Paris have become effectively contested and thus come to matter, as a way of navigating possibilities and constraints of change present in the substantial proportion of the urban fabric which lies beyond the control or influence of the main policy actors. Different bits or components of socio-technical systems become politicized in distinct ways and at particular times, whether it is contracts, production plants, pipes, steam or smart meters. I have attempted to distinguish between some of the processes and knowledge controversies at work in making/doing material politics, and their underlying notions of materiality as a form of relationship between ourselves and our physical world. Public transcripts of low carbon aims and efforts in official plans and strategies differ vastly from the everyday processes and practices of low carbon in people's work and lives. There is not just an implementation gap between what cities and organizations say they are doing or going to do and what they actually do, but also between what they publish and measure as objectives and areas of activity to achieve these and the less public, less recorded work, activities and 'events' which actually go on.

As we continue to explore the ongoing processes of urban transition, two issues which merit further reflection concern our approach to agencies of change and our need for stories depicting possible pathways. It may be productive to further explore a more distributed, relational and heterogeneous notion of agency – as situated in, around and between particular matters and materialities – in analysing activity, capacity and capability to effect low carbon/transition. Many of the artefacts and components tracked through the heat, electric and smart projects in Paris perform or do things in a way which often escapes or resists full control. Actors have at best a partial, temporary grasp over bits of very complex socio-technical systems such that notions of agency, flows and mechanisms of functioning and change are (increasingly) indistinguishable. At the same time, we are attached to political systems in which specific individuals have to be held accountable for actions or lack of actions (measured here, for example, by a declining curve on the

Paris emissions graph). Bennett (2005, p. 464) captures neatly the crucial dilemma we face: "It is ultimately a matter of political judgment what is more needed today: should we acknowledge the distributive quality of agency in order to address the power of human–non-human assemblages and to resist a politics of blame? Or should we persist with a strategic understatement of material agency in the hope of enhancing the accountability of specific humans?" There is no definitive response to this question, but sensitivity to the unruliness and vitality of matter (see Braun and Whatmore, 2010) may at the very least open up more of a critical hybridity perspective. This recognizes that the material politics of urban transition works through, and not against, human agency, but a recombinant agency in which the potentials and performances of materials and technologies can help to 'activate' new political actions and strategies (White et al., 2016).

It may also be useful to open up this matter and materiality to virtuous discourses and dreams which shape and are shaped by concrete enactment of urban transition, suggesting a need for more exploration of the role of narratives, fictions and fantasies in (re)materializing the multiple processes of transition and imagining and envisioning very real shifting relations between people and objects and new ways in which things are understood, used, circulated and experienced. In this regard, French theorist Armand Mattelart (1999, p. 178) identifies the 'thaumaturgic' virtues of infrastructure networks through history, and "the gap between prophecies based on the democratic potential of networks and the trajectory of realpolitik in their establishment." What this chapter has begun to reflect on, by exploring the process of urban (low carbon) transition through the effective materialities of heat, electric and smart infrastructure reconfiguration, is a bridging of this kind of gap. By focusing on always ongoing enrolments, deployments and disputes around infrastructure – the operational processes and politics through which transition comes to matter in the city – we develop a more open understanding of socio-technical change. Through this, we might begin to have faith in and forge new 'reticular utopias' (Mattelart, 1999) grounded in what matters to people and to their everyday struggles for social and ecological reproduction of meaningful urban living spaces.

Notes

1 As one interviewee put it for example: "We would like the term 'energy transition' to disappear and to be replaced by 'climate, air, energy'. At least that would be clear, because in my opinion in two years the energy transition will not be fashionable and everyone will have forgotten what it is" (City of Paris official, interview).
2 Local electricity supply is also a separate public service. The City of Paris has, however, attributed a single public service concession for both services (distribution and supply) to ERDF and EDF (as separate but closely linked companies). For the Regional Court of Accounts, this is "a source of confusion and opacity" (CRC d'Île-de-France, 2010, p. 2).
3 These administrative tribunals have the task of overseeing and verifying the accounts of public authorities and local governments.
4 A 2009 audit identified a level of underinvestment in the Paris network to the tune of between 750 million and 1 billion EUR, so it is an ageing infrastructure which has been 60% amortized (compared to a national average of 39%) (Baupin and Gassin, 2009).

5 It is only since the end of 2013 that ERDF has been obliged by State Council (Conseil d'État), the administrative supreme court, to provide detailed disaggregated technical and financial information about each of the contracts that it runs.

References

APUR (Atelier Parisien d'Urbanisme) (2013) *Une plateforme pour un PLU thermique (séminaire 10 juillet 2013)*. Paris: APUR.

Barry, A. (2013) *Material Politics: Disputes along the Pipeline*. Chichester: John Wiley & Sons.

Barry, A. (2015) Thermodynamics, matter, politics. *Distinktion: Journal of Social Theory*, 16(1), 110–125.

Baupin, D. and Gassin, H. (2009) *Concessions de distribution électrique à ERDF: retours sur l'expérience parisienne*. Paris: Mairie de Paris.

Beltran, A. (2002) *La ville-lumière et la fée électricité – l'énergie électrique dans la région parisienne: service public et entreprises privées*. Paris: Editions Rive Droite/Institut d'Histoire de l'Industrie.

Bennett, J. (2005) The agency of assemblages and the North American blackout. *Public Culture*, 17(3), 445–465.

Bezat, J.-M. (2010) La chambre régionale des comptes reproche à EDF d'avoir sous-investi dans la capitale. *Le Monde*, 28 September.

Braun, B. and Whatmore, S. (eds.) (2010) *Political Matter: Technoscience, Democracy, and Public Life*. Minneapolis: University of Minnesota Press.

Bulkeley, H., Castán Broto, V., Hodson, M. and Marvin, S. (eds.) (2011) *Cities and Low Carbon Transitions*. London: Routledge.

Cour des Comptes (2013) *Rapport public annuel 2013 – Les concessions de distribution d'électricité: une organisation à simplifier, des investissements à financer*. Paris: Cour des Comptes.

CRC d'Île-de-France (Chambre Régionale des Comptes d'Île-de-France) (2010) *Rapport d'observations définitives – Ville de Paris: Délégation du service public de distribution de l'énergie électrique dans Paris, exercices 2003 et suivants*. Paris: CRC d'Île-de-France.

CRC d'Île-de-France and PR d'Île-de-France (Conseil Régional d'Île-de-France and Préfet de la Région d'Île-de-France) (2012) *Schéma Régional du Climat, de l'Air et de l'Energie (SRCAE) de l'Île-de-France*. Paris: CRC d'Île-de-France and PR d'Île-de-France.

Energie2007.fr (2015) Available at: http://www.energie2007.fr/actualites/fiche/2269 (accessed 27 August 2015).

ERDF (Électricité Réseau Distribution France) (2011) *Repères et enjeux de la distribution d'électricité: dialogue avec ERDF* (seminar of 8 June). Paris: ERDF.

FNCCR (Fédération Nationale des Collectivités Concédantes et Régies) (2015) *Guide l'élu local et intercommunal: énergies*. Paris: FNCCR.

Ged, A. (2015) *Perspectives on Transition Pathways in Paris*. Paris: Agence Parisienne du Climat.

IssyGrid Project (n.d.). *IssyGrid*. Available at: http://issygrid.com (accessed 27 August 2015).

Joyce, P. and Bennett, T. (2010) Material powers: introduction. In: Bennett, T. and Joyce, P. (eds.) *Material Powers: Cultural Studies, History and the Material Turn*. London: Routledge, pp. 1–21.

Latham, A., McCormack, D., McNamara, K. and McNeill, D. (2009) Constructions. In: Latham, A., McCormack, D., McNamara, K. and McNeill, D. (eds.) *Key Concepts in Urban Geography*. London: Sage, pp. 53–87.

MacLeod, G. and Jones, M. (2011) Renewing urban politics. *Urban Studies*, 48(12), 2443–2472.

Mairie de Paris (2012a) *Bilan du Plan Climat 2007–2012*. Paris: Mairie de Paris.

Mairie de Paris (2012b) *Plan Climat Energie de Paris: grandes orientations*. Paris: Mairie de Paris.

Mattelart, A. (1999) Mapping modernity: utopia and communications networks. In: Cosgrove, D. (ed.) *Mappings*. London: Reaktion Books, pp. 169–192.

Rutherford, J. (2014) The vicissitudes of energy and climate policy in Stockholm: politics, materiality and transition. *Urban Studies*, 51(7), 1449–1470.

Ville d'Issy-les-Moulineaux (n.d.). IssyGrid: 1er réseau de quartier intelligent en France. Available at: http://www.issy.com/issygrid (accessed 27 August 2015).

Ville de Paris, ERDF and EDF (Ville de Paris, Électricité Réseau Distribution France and Électricité de France) (2009) *Avenant n°6 au traité de concession du 30 juillet 1955 pour la distribution de l'énergie électrique dans Paris*. Paris: Ville de Paris/ERDF/EDF.

White, D., Rudy, A. and Gareau, B. (2016) *Environments, Natures and Social Theory: Towards a Critical Hybridity*. London: Palgrave.

4

LEGACIES OF ENERGY AUTARKY FOR LOW CARBON URBAN TRANSITIONS

A comparison of Berlin and Hong Kong

Timothy Moss and Maria Francesch-Huidobro

1 Introduction

While cities are widely regarded as playing a pivotal role in energy transitions, today and in the future, recent research is highlighting the variety in urban responses to climate change (Bulkeley et al., 2011; Rutherford and Coutard, 2014; Mai and Francesch-Huidobro, 2015). The search for model development trajectories or institutional arrangements for the low carbon city is being clouded by stories of deviation, contestation, appropriation and adaptation peculiar to specific urban contexts and contingent events. This differentiated picture of urban energy transitions as they are happening is helpfully opening up the debate to the multifarious factors shaping urban transitions and the challenges that emerge from them for both policy and research. What is in danger of getting lost in these powerfully 'presentist' narratives is a sense of where these diverse urban responses are coming from and how historical legacies of energy production and use are influencing (low carbon) options for today and the future.

This chapter uses a comparative analysis of two iconic 'electric cities' – Berlin and Hong Kong – to explore the legacies of past socio-technical configurations for today's attempts to realign urban energy systems. The selection of these two cities is informed by their symbolic status as pioneers of the modern electrified city. Beyond their global symbolism as 'electric cities' the two cities are distinctive because of their unusual histories of autarky of power generation. Both cities have a long experience of being self-sufficient for their own power supplies and having to (re)configure their electricity systems around their own urban territory. For primarily geopolitical reasons, West Berlin and Hong Kong sought to secure their power supplies by maximizing urban energy autarky and limiting dependency on their regional neighbours: East Germany and mainland China. Since reunification in the 1990s, Berlin has had to realign its electricity system to take account of the reopening of borders, the introduction of competition and processes of economic and political integration with surrounding regions, while for Hong Kong reunification with China has made grid connections to, and electricity imports from, China the subject of growing debate (Environment Bureau, 2014, 2015). These energy transitions from urban autarky to regional integration – in the form of rapid realignment (Berlin) and gradual rapprochement (Hong Kong) – are accompanied by policy objectives to reduce carbon emissions, minimize energy use and increase energy efficiency through shifts in electricity generation and use.

The energy histories of these two cities inspire the following research questions, to be addressed in this chapter:

- Firstly, how did West Berlin and Hong Kong strive to render their electricity supply systems more autarkic in response to their geopolitical isolation and to what effect?
- Secondly, how have the two cities been realigning their electricity supply systems following reunification in the 1990s?
- Thirdly, how far and in what ways are their historical legacies of energy autarky framing options for energy transitions today?

This chapter is conceived not merely as a comparative case study of two 'electric cities' and their historical legacies, but also as a contribution to broader debates on energy autarky, energy security and urban energy transitions. It explores the impact of geopolitics on urban energy systems and the energy security strategies that have emerged in response to infrastructure isolation and subsequent integration. It highlights the spatial dimensions of these strategies, as exemplified in the reconfiguration of electricity networks around the territorial confines of the city. It also advances knowledge on the history of urban energy by illustrating how far and in what ways past events and trajectories influence today's energy systems and their prospects for transition.

2 Urban energy transitions: discourses of autarky and security

Recent contributions on urban energy transitions by human geographers, political scientists and sociologists have raised awareness of how local contexts are shaping place-specific transitions in practice (Bulkeley et al., 2011; Rutherford and Coutard, 2014; Mai and Francesch-Huidobro, 2015). From this corpus of work we have learned, for instance, that energy transitions in cities do not follow a model or linear pathway; they are often highly contested and they tend to overlay, rather than replace, existing modes of energy provision. This literature can, however, be criticized for its strongly 'presentist' perspective on urban energy transitions, focusing on current attempts to promote low carbon cities and relegating the historical legacies of urban energy systems to introductory contextualization. This research deficit is met to some extent by scholars of urban environmental history and the history of technology who have explored earlier energy transitions, for instance from wood to coal, from gas to electricity or from municipal to national power utilities (Hughes, 1983; Melosi, 2000). What is still missing, though, is research spanning these two bodies of literature, i.e. studies capable of explaining how the history of a city's energy system is influencing today's energy transitions.

In this chapter we offer an illustration of this research potential by setting the ongoing energy transitions in Berlin and Hong Kong in the context of their recent urban energy histories. What makes this endeavour intriguing is that both cities are adapting to a very different kind of transition to their energy systems; namely, the reintegration of their insular urban networks into regional and national electricity systems. In the case of Berlin, technical networks have been reconnected, new organizational structures created, regulatory regimes altered and resource flows redirected. In the case of Hong Kong, such processes of infrastructural and market integration are ongoing. Yet, at the same time, elements of the old autarkic electricity systems still remain entrenched. This offers an excellent opportunity to study processes of reconfiguration of urban energy networks in a city's recent history, observing which elements change and which do not. It also allows us to investigate not just one energy transition in each city, but two, exploring how today's attempt to pursue a low carbon agenda for each city is constrained or assisted by its legacy of autarky, concerns over energy security and steps towards spatial reintegration.

These issues of energy autarky and energy security in connection with shifting energy geographies resonate with several strands of recent academic debate on (urban) energy transitions. Energy autarky, as discussed in the context of the current low carbon and energy transitions debate, is conceived most frequently in the normative sense of a programmatic vision (Müller et al., 2011). Müller at al. define energy autarky not simply in terms of self-sufficiency of supply but also regarding the energy source (e.g. renewables from the region, rather than carbon energy imports), the decentralized structure of the energy system and increased energy efficiency on both the supply and demand side (2011, p. 5802). This literature does not reflect critically on the assumptions underpinning the connectivity between autarky, renewables and decentralized organization, nor does it consider potential drawbacks of energy autarky, such as negative impacts on the surrounding region or issues of legitimacy emerging from community initiatives (Hodson and Marvin, 2010). In this chapter we use the term energy autarky in a simple, non-normative sense to refer to a policy of establishing self-sufficiency in electricity generation (rather than full energy autonomy) in a specific territory – in our case, two isolated cities.

Building autarky into an electricity supply system is one strategy to increase energy security in cities, the second strand of literature of relevance to this chapter. In their work on urban ecological security Hodson and Marvin draw attention to recent trends of cities and their utilities to make their socio-technical networks more resilient to shocks and stresses (Hodson and Marvin, 2010; see also Medd and Marvin, 2005). They illustrate how energy autarky is becoming a means of increasing urban resilience with the examples of New York City's strategy of energy self-sufficiency and London's ambitious targets for decentralized energy

(Hodson and Marvin, 2010). The emergent strategy of urban ecological security they identify is about reconfiguring infrastructures to safeguard resource flows and essential services in the face of a growing variety of threats, ranging from impacts of climate change to terrorism and warfare.

This chapter contributes to these two bodies of work by exploring connectivity between energy autarky and energy security as components of urban energy transitions. It investigates two past instances where energy autarky became regarded as a political and technical necessity to secure the continuous and adequate supply of electricity in each city. We aim to show for each case how this came to be, what form energy autarky took, how this is being reordered following reunification and what impacts this historical legacy is having on today's attempts at transition towards a low carbon city.

3 Urban energy autarky as a response to geopolitical isolation

West Berlin, 1948–90

On 24 June 1948 the Soviets instructed Berlin's electricity utility Bewag, located in the Soviet sector, to cut off all power supplies to the three western sectors of the city (Moss, 2009). Deliveries of coal to power plants in the western part were also to cease. This sudden truncation of flows of electricity and coal plunged West Berlin into a supply crisis far more severe than anything experienced during the wartime bombardment and invasion of the city. The immediate response of the Western Allies revealed how vulnerable West Berlin was to the power cut-off. The airlift set up to overcome the blockade transported not so much food and clothing as coal needed to keep West Berlin's power stations in operation. Indeed, whole generators were flown in for the reconstruction of the main power station Kraftwerk West (Senat von Berlin, 1964, p. 736). Notwithstanding these emergency measures electricity use had to be restricted to an average two hours during the day and two hours at night during the winter of 1948–49. Per capita consumption in West Berlin fell to a mere quarter of East Berlin levels. Parallel to the physical division of the infrastructure networks the Berlin blockade heralded also the organizational split of the power utility Bewag. Following a lengthy struggle for control of the company it was divided into a West Berlin and East Berlin utility in December 1948 (Merritt, 1986).

The end of the blockade on 12 May 1949 did not restore external electricity supplies to West Berlin. Despite a number of agreements on the delivery of electricity from East Berlin over the following years, none of these proved reliable. In March 1952 Bewag-East precipitated the termination of all electricity supplies to West Berlin (Merritt, 1986). From then until German reunification in 1990 West Berlin was to remain an 'electricity island', cut off from the national and regional grids of both West and East Germany (Merritt, 1986; Varchmin and Schubert, 1988; Bewag, 1991).

It is against this prolonged experience of system vulnerability that West Berlin's subsequent strategy of expanding its electricity generating capacity needs to be

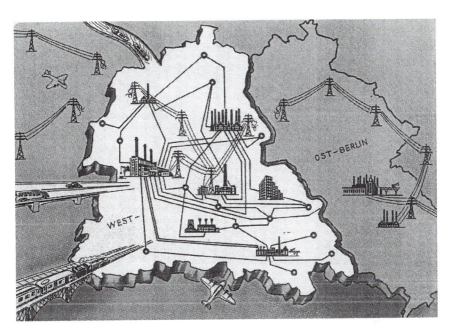

FIGURE 4.1 West Berlin's truncated electricity system, 1956.
Source: Stolpe, 1956, p. 277.

understood. The priority was to reduce dependency on East Berlin and the Soviet zone by achieving full energy autarky – at least in terms of power generation (Stolpe, 1953, 1956). By 1952, Kraftwerk West had been restored to a capacity of 268 megawatts (MW), already providing the lion's share of the total 382 MW at West Berlin's disposal (Brocke and Brüss, 1953). Major extensions were made to other power stations in the boroughs of Moabit, Charlottenburg, Steglitz and Spandau. By early 1955, Bewag-West had become self-sufficient in electricity generation, largely thanks to credits to fund capital investments from the European Recovery Programme (ERP). By 1984, West Berlin boasted a generating capacity of 2,251 MW, capable of covering the city's electricity demand that year, standing at 8,990 mill kWh (Bers and Strempel, 1984; Ziesing, 1985). Throughout the period of political division the city remained entirely dependent on shipments of coal (and, increasingly, oil) from outside to run the power stations, but strove to minimize vulnerability to disruption by maintaining huge coal reserves within the city.

There is no indication that Bewag-West considered saving energy and curbing demand as a way of helping secure electricity supply in its insular situation. It defied political pressure to promote demand-side management and, subsequently, renewable energies (Monstadt, 2007; Moss, 2014). This is all the more surprising since the utility, to meet anticipated growth in demand, built up reserves of generating capacity in excess of conventional levels and had to rely on expensive imported fuel. The strategy was energy security via energy autarky, to be achieved primarily by expanding generating capacity within the city. It took a number of protests against the construction of new power stations, together with the oil crisis in 1973, for Bewag-West to reconsider its expansionist strategy, albeit in a modest way (Monstadt, 2004). A radical policy change was heralded only in 1989 with the election of the city's first social democratic–green coalition, which created an energy task force to promote more efficient power stations, renewable energies and demand-side management and, in 1990, a new state Energy Law setting out an institutional framework for energy planning (SenStadtUm, 1990; Monstadt, 2007). By then, however, the fall of the Berlin Wall and the reunification of the city had created a new setting for energy policy in Berlin.

Colonial Hong Kong, 1841–1997

Unlike West Berlin, Hong Kong Island had been self-sufficient in energy generation for electricity since power was first introduced to the island in the late 1800s and was to remain so throughout British rule (1841–1997). The first power utility, the Hong Kong Electric Company Ltd (HEC), was established in 1890 and built a succession of coal-fired power stations, beginning with the Wanchai Power Station in 1890 and followed by North Point A (1919) and B (1958), both decommissioned in 1989. Two oil-fired power stations were built in 1966 and 1968 at North Point C and Ap Lei Chau, on the south side. These were also decommissioned in 1989 to make space for homes (Hong Kong Electric, 2014; Waters, 1990). HEC began relocating power generation operations at this time to Lamma Island (off the south coast of Hong Kong Island), where it constructed eight coal-fired generators and five gas turbines in operation from 1991 to 1997. In return for providing a high level of service, HEC was granted a secure source of funding for investments, as set out in a Scheme of Control Agreements (SCA), in force from 1979 to 1992 and subsequently renewed (Hong Kong Electric, 2014).

In the Kowloon Peninsula and the New Territories, the history of electricity began in 1901, when the China Light and Power Company Syndicate (later CLP Power Hong Kong Ltd) was formed (CLP Power Ltd, 2014; Clifford, 2015). The first CLP power plant (oil-fired) was opened in 1903 at Chatham Road with a generating capacity of 75 kW, providing street lighting from 1919 onwards. It consolidated its generating capacity during the 1950s and 1960s, purchasing the Tai O (Lantau Island) Power Company in 1955 to supply Lantau Island and commissioning another oil-fired power station at Tsing Yi in 1969 (CLP Power Ltd, 2014). Like HEC, CLP concluded an SCA with the Hong Kong government for the period 1978–92, followed by a second, covering the period 1993–2008. The reliance on oil ended following the oil crisis of 1973, with the commissioning of the coal-fired power stations Castle Peak A (1979) and Castle Peak B (1986). Subsequently, gas-fired power stations were built by the company at Penny's Bay in the eastern part of Lantau (1992) and at Black Point (1996).

During the 1980s CLP began to look beyond the British colony and take initial steps towards connecting the Hong Kong power system to that of the Chinese mainland. It had started selling electricity to Guangdong Province as early as 1979.

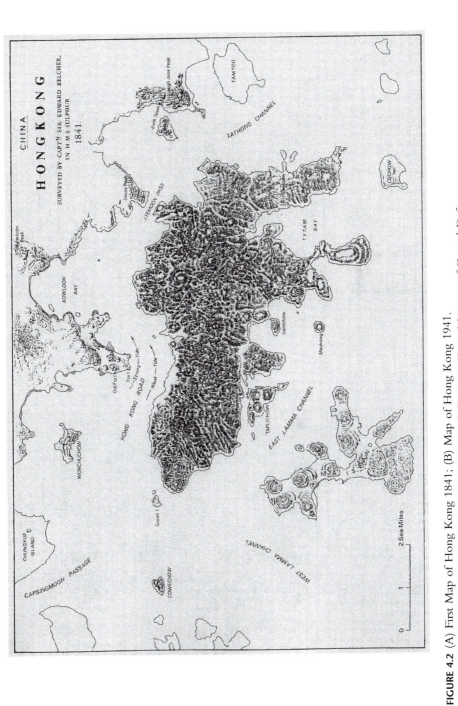

FIGURE 4.2 (A) First Map of Hong Kong 1841; (B) Map of Hong Kong 1941.
Source: photos by Francesch-Huidobro of the exhibition hall, Hong Kong Museum of Coastal Defence.

PLAN TO ACCOMPANY REPORT
BY SIR DAVID J. OWEN
ON THE FUTURE CONTROL AND
DEVELOPMENT OF THE
PORT OF HONG KONG
FEBRUARY 1941.

FIGURE 4.2 (continued)

In 1985 CLP signed the Daya Bay Joint Venture with the Chinese government. On the basis of this agreement, CLP – together with Guangdong Nuclear Investment Company – commenced construction of a pressurized water reactor nuclear power station in Daya Bay, Guangdong Province, commissioned in 1994 (CLP Power Ltd, 2014). Also in 1985, CLP adopted a business model including a commitment to develop wind and hydro power in mainland China (Chow, 2001).

In sum, Hong Kong's colonial period was characterized by a stable electricity regime built up around the two territorial monopolies of HEC and CLP. Geopolitical isolation encouraged both to pursue a strategy of safety and reliability of supply, safeguarding resource flows and maintaining affordable services within the city. Until the Daya Bay nuclear plant was commissioned in the mid-1990s, Hong Kong was entirely self-sufficient in its power supply. Its dependence on external fuel sources did, however, prompt shifts during the century, notably from oil to gas and nuclear following the oil crisis.

4 Reintegrating infrastructure in response to reunification

Berlin, 1990–present

The reunification of Germany in 1990 brought to an abrupt end West Berlin's status as an 'electricity island'. After forty years of division caused by geopolitical isolation the immediate tasks were to reconnect the power supply systems of East and West Berlin, amalgamate the two power utilities and link the West Berlin electricity network to the national and European grids (Bewag, 1991; Winje, 1994; Monstadt, 2004). Overcoming these technical and organizational ruptures proved relatively straightforward over the following years. In 1990 the power utility of East Berlin was amalgamated into West Berlin's Bewag. A permanent 380 kilovolt cable connecting West Berlin to the national grid followed in 1994.

What proved far harder for Berlin's electricity system post-1990 was adapting to an energy market and policy context which was becoming increasingly competitive, diverse and environmentally sensitive (on the following Monstadt, 2007, pp. 330–335). The initial economic impact of German unification on Berlin was to precipitate the collapse of its traditional industries, a fiscal crisis in the early 1990s and the sale of assets. Most of the city's utilities were sold, including its majority shareholding in Bewag in 1997. The following year, 1998, Germany's electricity market was liberalized in accordance with European legislation, requiring the freshly privatized Bewag to compete for business. Without its territorial monopoly for power supply in Berlin, the utility sought to remain competitive by cutting costs, primarily in investments to the network, but also in environmental protection, R&D activities and consumer services (ibid.).

It is in this context of market liberalization and cost-cutting that the power generating capacity built to keep West Berlin powered suddenly became a commercial liability. The inner-city power stations constructed in the post-war years had been designed to serve energy security interests, not low-cost electricity provision, and many proved uncompetitive in the new national (and European) electricity market. Consequently, seven power plants in the former West Berlin, with a total generating capacity of 1,884 MW, were closed down between 1999 and 2005 (SenWTF, 2011, p. 45). This had the positive side effect of reducing CO_2 emissions within the city significantly, which between 1990 and 2011 fell by nearly a half, from 13.385 to 6.747 million metric tons (AFS, 2014, p. 33). However, the downside of decommissioning power stations has been to reduce substantially the city's district heating capability, since most of them produced heat as well as power. As Berlin dismantles its overcapacity in unprofitable coal-fired electricity generation, the city is becoming increasingly reliant on power from outside the city, primarily from lignite-fired power stations in the neighbouring region of Lusatia. This has been raising criticism that the city is seeking to meet its CO_2 reduction targets by importing a growing proportion of its electricity from outside, disregarding the impacts on the climate. The proportion of electricity generated in the city using renewables is, by national comparison, low, at 7.0% in 2011 (ibid., p. 27).

In terms of energy policy, the conservative–social democratic city government formed in 1990 pursued the path set out by its red–green predecessor, but with far less enthusiasm. The Energy Concept for Berlin of 1994 targeted a 25% reduction in CO_2 emissions between 1990 and 2010 and included an energy-saving action plan, covering housing, public buildings, transport, energy services and renewable energies (Monstadt, 2004). Responsiveness to climate change became, for the first time, an issue of energy security. In the early 1990s new agencies were created, such as the Berlin Energy Agency and the Energy Coordinating Unit within the Senate Department for Urban Development and the Environment, which provided important groundwork for today's energy transition in the city. However, the city government was not prepared to force Bewag to prioritize energy efficiency, especially not following its full privatization in 1997. In the absence of regulatory pressure on its own utility, the Berlin government resorted to public funding for projects planned under the Energy Concept. As the city's debt spiralled from 1992 onwards, funding for these projects dropped sharply. Henceforth, the city relied increasingly on private contractors to implement its energy-saving programmes – to limited effect.

In 2010, under a social democratic–socialist party coalition, the city government adopted a new Energy Concept targeting a 40% reduction in CO_2 emissions by 2020 and promoting greater energy efficiency and renewable energies (SenWTF, 2011). The experience of Berlin's energy policy following reunification (Monstadt, 2004) cautions, however, against expectations of a strategic shift in direction. Beyond state policy, though, there are interesting recent initiatives for the city or its residents to take over ownership of the city's power grid, following the expiry of the latest concession, and to set up a new municipal power utility which is more environmentally sustainable, democratically accountable and socially responsible than the incumbent Vattenfall (Moss et al., 2015).

Hong Kong, 1997–present

The return of Hong Kong to Chinese sovereignty on 30 June 1997 did not unleash a process of physical and institutional integration of two energy systems as witnessed in Berlin. Unlike the case of Berlin, Hong Kong's electricity system had been, from its early beginnings until only recently, completely separate from that of mainland China. Any form of integration between the Hong Kong and Chinese networks following reunification was, therefore, not about restoring a former status of connectivity, but about developing new modes of interaction under the constitutional principle of "one country, two systems" (Hong Kong SAR Government, 1991).

At first glance, little appears to have changed in Hong Kong's power system since 1997. The two incumbent power utilities, HEC and CLP, continue to own and operate the grids and supply electricity in their respective areas of the city. The strategy of both utilities to build up capacity to ensure reliability of supply is enshrined in the SCAs agreed with successive governments from colonial times to the present. Mechanisms in the SCAs, such as the permitted rate of return based on average fixed assets, encourage investment in capital works dependent on the amount of electricity sold, thus stimulating, rather than minimizing, demand. Hong Kong continues to consume high levels of electricity. Fossil fuels still dominate the fuel mix for power generation in Hong Kong, with coal accounting for 53% in 2012, followed by imported nuclear energy from mainland China (23%), natural gas (22%) and renewables (2%) (Censtat, 2013).

FIGURE 4.3 Daya Bay Nuclear Power Plant, Canton, China, 2012.
Source: photo by M. Francesch-Huidobro.

A closer look, however, reveals that Hong Kong's tradition of autarky in power generation and supply is gradually being undermined. New developments since the return of sovereignty to China are marked by the growing prominence of cross-border electricity transfers, the opening up to contestable energy markets, policy initiatives to mitigate climate change and public pressure for greater demand-side management. These developments are, in very different ways, challenging the Hong Kong government and the two city utilities to give up the traditional policy of energy autarky through network expansion and to seek greater regional integration.

An early indication of this shift was the public consultation on an extended SCA for CLP and HEC conducted in 2005. The recommendations emerging from it included the creation of an independent regulatory authority to supervise a gradual liberalization of the electricity market after 2008, favouring access to providers of renewable energy, and a reduction in the rate-of-return driving network expansion (Francesch-Huidobro, 2014a, 2014b). Intriguingly in terms of energy autarky, the recommendations also required HEC and CLP to purchase electricity from mainland China when available and cheaper than local production (EDLB, 2005). A Memorandum of Understanding was signed between Hong Kong and China in 2008 to promote energy cooperation, specifically over the continuous supply of nuclear power and natural gas from China.

A second driver of change has been climate policy. Following its obligation to meet the Chinese government's CO_2 reduction targets set in 2009, the Hong Kong government has committed itself to reducing greenhouse gas emissions by 50–60% by 2020, in excess of the national target of 40–45%. This commitment was subsequently specified in the *Hong Kong Climate Change Strategy and Action Agenda 2010* (EPD, 2010). The strategy proposes a shift in Hong Kong's electricity generation fuel mix, increasing the shares of natural gas to around 40% and of nuclear power provided from mainland China to 50% by 2020 (EPD, 2010, p. 43). Despite its high ambitions for climate mitigation, this policy has been widely criticized by local environmental non-governmental organizations (NGOs) not only because of safety concerns over nuclear power but, more importantly, because of the absence of demand-side management measures to reduce peak electricity demand and end-user energy consumption (Francesch-Huidobro et al., 2014).

Two further consultation exercises conducted in 2014 and 2015 revealed that public opinion in Hong Kong favours using more locally generated electricity from natural gas, coal and renewable energy over importing more electricity from the mainland China power grid (Environment Bureau, 2014, 2015). Serious concerns are being voiced by user groups, political parties, environmental NGOs and research organizations not only about the reliability of electricity supplied from mainland China (Wan, 2014), but also about the time required to build the cross-border interconnector (CLP Power Ltd, 2014). Particularly contested is the creation of a Regional Transmission Organization in the Pearl River Delta Region (Hong Kong's hinterland), which may be used to advance a unified grid and power services between Hong Kong and the mainland.

The option for more locally generated power is, however, also not free of problems. Increasing the share of natural gas to 60%, as envisaged for this option, would create new dependencies on the West-East Gas Pipeline from Turkmenistan, following the depletion of gas resources from the South China Sea gas field Yacheng 13-1 off Hainan Island, as well as on liquefied natural gas sourced from Shenzhen, Guangdong Province. Furthermore, the expansion of renewable energy sources for Hong Kong is severely limited by its physical geography. High urban density and a lack of land restrict the use of solar and wind power on the territory. HEC and CLP are currently exploring the feasibility of building large-scale offshore wind farms with a total installed capacity of 300 MW, but even this would cover only 1.5% of Hong Kong's electricity consumption (CLP Power Ltd, 2014). Given these physical, as well as political, limitations to change, the incumbent utilities point to their past track record of high service reliability and relatively low tariffs as a strong argument for retaining the status quo (CLP Power Ltd, 2015).

5 Legacies of local autarky and regional realignment for energy transitions today

Having explored in the previous sections how each of the two 'electric cities' developed their own forms of energy autarky in response to their geopolitical isolation and how they are currently realigning their electricity systems to fit conditions post-reunification, we now turn to our third research question and consider how these historical legacies are framing options for energy transitions today. The purpose of this section is to discuss what we have presented in terms of how the past is influencing the present and future options.

Looking across the two cases, the first observation to make is that the experiences of energy autarky made during and after periods of geopolitical isolation do not fit into neat categories of isolation versus integration. The energy pathways pursued by each city may have been powerfully influenced by the desire to seek security in autarky of electricity generation, but they were never completely independent of the outside world, as the reliance on external fuel sources and latter-day grid connections to the hinterland show. Conversely, political reunification did not cause the legacies of energy autarky to lose their relevance overnight. They continue to influence current policies and practices of energy provision and use; not in a deterministic and overpowering way, but in a subtler and selective manner. This may be an obvious point, but it is by no means trivial. It guides us to analyse, in this section, the extent to which these historical legacies are framing today's urban energy systems.

1. *Legacies of territorial integrity*: During periods of geopolitical isolation both cities sought to develop autarky of electricity generation and supply to defend their territorial integrity. West Berlin's history of territorial protection – as a response to forced separation – is today largely insignificant in the reunited city. In Hong Kong, by contrast, the legacy of territorial integrity built up by

the British and now enshrined in the SAR Basic Law remains a key issue in terms of relations with mainland China. Ongoing debates on closer cooperation with China over Hong Kong's future electricity supply are framed by fears of Hong Kong losing its ability to pursue its own policy agenda.

2. *Legacies of protected markets*: The local power utilities in West Berlin and Hong Kong were instrumental in maintaining territorial integrity by providing reliable energy services. To this end they were protected from competition and allowed to operate territorial monopolies. The separation of Bewag in Berlin from the national grid during the years of political division and the self-dependence this cultivated made the transition to a liberalized electricity market after 1998 especially hard, resulting in cost-cutting and plant closures to remain competitive. In Hong Kong, the SCAs agreed between the government and the two power utilities were designed to protect HEC and CLP from market competition in return for securing a high and reliable level of service. Since successive SCAs have been modified in only a modest way, Hong Kong's electricity system remains characterized, essentially, by two vertically integrated territorial monopolies.

3. *Legacies of supply security*: Securing electricity supplies was the guiding principle of energy policy in both cities during their geopolitical isolation. To allow for potential disruption of service huge capacities for generation were built up in each city, in the shape of multiple power stations and grid extensions. It is interesting to note that neither city pursued the option of minimizing electricity consumption and, thereby, reducing the need to expand capacity. The legacy in both cities is continued reliance on the logic of build-and-supply to the detriment of demand-side management. Establishing energy efficiency on the policy agenda is, for this reason, proving difficult in Berlin and Hong Kong.

4. *Legacies of local infrastructure*: The construction of sufficient power stations within its own territory posed a particular challenge in both cities. Especially in Hong Kong, competing claims on land for residential and business purposes caused inner-city power plants to be replaced by ones at increasingly peripheral locations. In Berlin after 1990, the problem lay rather in how to deal with the legacy of its urban power stations in a liberalized energy market. Designed during the years of division to maximize generation capacity, these power stations were ill-equipped to compete, proving a commercial liability to the utility Bewag. Several were consequently decommissioned during the 2000s, but at the price of losing a substantial proportion of Berlin's district heating capacity.

5. *Legacies of resource flows*: Energy autarky in West Berlin and Hong Kong was always only limited to electricity generation, as both were dependent almost entirely on fuel imports. It is interesting to observe a gradual process of diversification of fuel sources over the years from coal to more oil and gas – also largely for geopolitical reasons – in particular in Hong Kong. This heavy dependence on fossil fuels is today creating difficulties for both cities in achieving the ambitious CO_2 reduction targets they have set themselves.

Pressure to deliver on climate mitigation goals is tempting both city governments to reduce local generation and import more electricity. This practice is criticized by environmental groups for concealing the fact that the CO_2 emissions are merely being produced elsewhere.

6. *Legacies of environmentalism*: Both West Berlin and Hong Kong have – relative to their respective hinterlands – a strong tradition of local environmentalist movements, arising partly in opposition to the dominant energy strategy of their governments and utilities. The alternative cultures of both cities, attracting especially young people interested in unconventional lifestyles and a more open society, have also nurtured support for green politics. These protagonists of sustainable forms of energy provision and use are playing a crucial role in ensuring that the energy transition policies of their respective cities are not the preserve of the political and business elites.

6 Conclusions

The purpose of this chapter has been to draw attention to the ways in which today's urban responses to climate change and energy security are framed by historical legacies of local energy systems and their embeddedness in – or disconnection from – wider regional, national and transnational structures. To illustrate the non-linearity and contestation of urban energy pathways we deliberately chose for study two cities – Berlin and Hong Kong – whose current attempts to address climate change are, we argue, shaped by their histories of self-reliance in power generation. The legacy of energy autarky lives on, in very different ways, in both cities, creating particular challenges for realigning their energy systems to meet new requirements of climate change, market competition, energy security and regional cooperation. Realigning the 'electric city' effectively means taking its history seriously.

What, though, is the significance of this study for current debates on energy autarky and energy security and their role in urban energy transitions? There are obvious limitations to generalizing from the experience of such exceptional, iconic cases. The value of studying West Berlin and Hong Kong lies not in any model status they may be ascribed, but in what their unusual histories can reveal about urban energy transitions that is absent or less visible in other cities with more conventional trajectories. In terms of energy autarky and energy security the stories of Berlin and Hong Kong make very clear that these concepts have not always been interpreted in the ways they are today and that their meanings can be very different in particular spatial-temporal contexts. The energy autarky pursued in West Berlin after the blockade and throughout colonial Hong Kong did not follow today's normatively loaded understanding of an entity aspiring voluntarily to local self-sufficiency based on renewable sources, low demand and citizen participation. For these two cities it was, rather, imposed upon them by geopolitical circumstances, and involved securing electricity supply *without* having to change practices of consumption, technologies of generation or governance structures. This suggests we need to question assumptions about what energy autarky can be and what forms it can take. Our two cases suggest that it can be a geopolitical necessity and a mode of energy security, but also a restriction on future energy options and an inefficient use of energy resources. Similarly, energy security can mean different things at different times and in different places. Both Berlin and Hong Kong have to come to terms with rapidly shifting meanings of energy security. What was once principally about protecting their energy systems from hostile neighbouring states, today encompasses a variety of concerns ranging from climate protection and ecological security to fuel availability and terrorist attacks.

Future research on urban energy transitions would benefit from unpacking these notions of energy autarky and energy security in terms of how they are interpreted and used in energy policy and research. Beyond this, our study has argued that historical legacies are hugely important to understanding urban energy transitions. Here, it is not enough to acknowledge that history matters. We need to reveal *how* history matters, for instance in the ways in which some components of an urban energy regime remain obdurate, some disappear or are discarded, while others adapt to shifting contexts or emerge in the wake of contingent events. The processes by which these socio-technical components get realigned – or reassembled – are often contested and non-linear. Although strongly influenced by the past, they are not determined by it. This points scholarship in the direction of more nuanced studies of socio-technical legacies, unpacking the dynamics and continuities of the "seamless web" and its multiple components (institutional, technical, geographical, social etc.). The two stories also sensitize us to the context-dependency of energy and low carbon policies. The strategy of both cities to build up huge capacities for power generation that elsewhere would be viewed as expensive, inflexible and unwise was, in their situation, essential to secure supply. Finally, our study warns against taking too myopic a view on cities when analysing urban energy transitions. The energy histories of Berlin and Hong Kong cannot be studied as purely urban phenomena: the pursuit of energy autarky was in direct response to their geopolitical situation; the realignment today is equally influenced by their relations with surrounding territorial entities. We need to acknowledge and explore the multiple ways in which urban energy transitions are shaped by, and themselves shape, regional, national and international developments.

Acknowledgements

A longer version of this chapter is published as Moss, T. and Francesch-Huidobro, M. (2016) 'Realigning the electric city: legacies of energy autarky in Berlin and Hong Kong', *Energy Research and Social Science*, 11, 225–236. The authors are grateful to the publishers for permission to reprint.

References

AFS (Amt für Statistik Berlin-Brandenburg) (2014) *Energie- und CO$_2$-Bilanz in Berlin 2011*. Statistical report EIV4-j/11. Berlin: Statistischer Bericht.

Bers, K. and Strempel, R. (1984) Stromerzeugung und Umweltschutz bei der Bewag im Wandel der Zeit. *Elektrizitätswirtschaft*, 83(9/10), 447–454.

Bewag (ed.) (1991) *Strom für Berlin: Von der Spaltung zur Wiedervereinigung*. Berlin: Berliner Kraft und Licht(Bewag)-Aktiengesellschaft.

Brocke, W. and Brüss, L. (1953) Die Berliner Energieversorgung. In: Institut für Raumforschung, Köln (ed.) *Die unzerstörbare Stadt: Die raumpolitische Lage und Bedeutung Berlins*. Berlin: Carl Heymanns, pp. 111–118.

Bulkeley, H., Castán Broto, V., Marvin, S. and Hodson, M. (2011) *Cities and Low Carbon Transitions*, London: Routledge.

Censtat (Census and Statistics Department) (2013) *Hong Kong Energy Statistics*. Hong Kong: Hong Kong SAR, Industrial Production Statistics Section, Census and Statistics Department.

Chow, C.H. (2001) Changes in fuel input of electricity sector in Hong Kong since 1982 and their implications. *Energy Policy*, 29(15), 1399–1410.

Clifford, M. (2015) 'Adhere and prosper': one company's quest for green power. Ch. 8 of: *The Greening of Asia: A Business Case for Solving Asia's Environmental Emergency*. New York: Columbia Business School Publishing.

CLP Power Ltd (2014) Our history. Available at: https://www.clp.com.hk/ourcompany/aboutus/ourhistory/Pages/ourhistory.aspx (accessed 1 February 2014).

CLP Power Ltd (2015) CLP's preliminary views on the consultation document on the future of the electricity market. Email communication, 31 March.

EDLB (Economic Development and Labour Bureau) (2005) Consultation paper on future development of the electricity market in Hong Kong: Stage I. Available at: http://www.enb.gov.hk/sites/default/files/en/node23/consultation_paper_stage2.pdf (accessed 1 January 2015).

Environment Bureau (2014) *Public Consultation on Planning Ahead for a Better Fuel Mix: Future Fuel Mix for Electricity Generation*. Hong Kong: Hong Kong SAR Government. Available at: http://www.gov.hk/en/residents/government/publication/consultation/docs/2014/FuelMix.pdf (accessed 1 August 2014).

Environment Bureau (2015) *Public Consultation on the Future Development of the Electricity Market*. Available at: http://www.enb.gov.hk/sites/default/files/en/node3428/EMR_condoc_e.pdf (accessed 12 August 2015).

EPD (Environmental Protection Department) (2010) *Hong Kong's Climate Change Strategy and Action Agenda*. Available at: http://www.epd.gov.hk/epd/sites/default/files/epd/english/climate_change/files/Climate_Change_Booklet_E.pdf (accessed 12 August 2015).

Francesch-Huidobro, M. (2014a) Climate policy learning and change in cities: the case of Hong Kong and its modest achievements. *Asia Pacific Journal of Public Administration*, 36(4), 283–300.

Francesch-Huidobro, M. (2014b) Public participation and sustainable development in Hong Kong. In: Cheng, Y.S. (ed.) *New Trends of Political Participation in Hong Kong*. Hong Kong: City University of Hong Kong Press, pp. 139–182.

Francesch-Huidobro, M., Tang, S.K. and Stratigaki, E. (2014) *A Carbon Reduction Implementation and Assessment Strategy for Hong Kong (CRiAS)*. Hong Kong: City University of Hong Kong/AECOM.

Hodson, M. and Marvin, S. (2010) *World Cities and Climate Change: Producing Urban Ecological Security*. Maidenhead: Open University Press.

Hong Kong Electric (2014) Electricity generation. Available at: http://www.hkelectric.com/web/AboutUs/Generation/Index_en.htm (webpage no longer accessible, accessed 1 December 2017).

Hong Kong SAR Government (1991) Basic Law. Hong Kong: Hong Kong SAR Government. Available at: http://www.basiclaw.gov.hk/en/index/index.html (accessed 1 January 2015).

Hughes, T.P. (1983) *Networks of Power: Electrification in Western Society, 1880–1930.* Baltimore, MD: Johns Hopkins University Press.

Mai, Q. and Francesch-Huidobro, M. (2015) *Climate Change Governance in Chinese Cities.* London: Routledge.

Medd, W. and Marvin, S. (2005) From the politics of urgency to the governance of preparedness: a research agenda on urban vulnerability. *Journal of Contingencies and Crisis Management*, 13(2), 44–49.

Melosi, M. (2000) *The Sanitary City: Urban Infrastructure in America from Colonial Times to the Present.* Baltimore, MD: Johns Hopkins University Press.

Merritt, R.L. (1986) Postwar Berlin: divided city. In: Francisco, R.A. and Merritt, R.L. (eds.) *Berlin between Two Worlds.* Boulder, CO: Westview Press, pp. 153–175.

Monstadt, J. (2004) *Die Modernisierung der Stromversorgung. Regionale Energie- und Klimapolitik im Liberalisierungs- und Privatisierungsprozess.* Wiesbaden: Verlag für Sozialwissenschaften.

Monstadt, J. (2007) Urban governance and the transition of energy systems: institutional change and shifting energy and climate policies in Berlin. *International Journal of Urban and Regional Research*, 31(2), 326–343.

Moss, T. (2009) Divided city, divided infrastructures. Securing energy and water services in post-war Berlin. *Journal of Urban History*, 35(7), 923–942.

Moss, T. (2014) Socio-technical change and the politics of urban infrastructure: managing energy in Berlin between dictatorship and democracy. *Urban Studies*, 51(7), 1432–1448.

Moss, T., Becker, S. and Naumann, N. (2015) Whose energy transition is it, anyway? Organisation and ownership of the Energiewende in villages, cities and regions. *Local Environment*, 20(12), 1547–1563.

Müller, M.O., Stampfli, A., Dold, U. and Hammer, T. (2011) Energy autarky: a conceptual framework for sustainable regional development. *Energy Policy*, 39(1), 5800–5810.

Rutherford, J. and Coutard, O. (2014) Urban energy transitions: places, processes and politics of socio-technical change. *Urban Studies*, 51(7), 1353–1377.

Senat von Berlin (1964) *Berlin: Quellen und Dokumente 1945–1951.* Berlin: Heinz Spitzing.

SenStadtUm (Senatsverwaltung für Stadtentwicklung und Umweltschutz) (ed.) (1990) *Ziele und Möglichkeiten einer stromspezifischen Energiepolitik in Berlin (West) unter Berücksichtigung des Stromverbundes mit der Bundesrepublik.* Berlin: Kulturbuch-Verlag.

SenWTF (Senatsverwaltung für Wirtschaft, Technologie und Frauen) (ed.) (2011) *Energiekonzept 2020.* Berlin: Langfassung.

Stolpe, R. (1953) Der Wiederaufbau der West-Berliner Stromversorgung. *Elektrizitätswirtschaft*, 52(14), 366–369.

Stolpe, R. (1956) Berlin – Großstadt ohne Verbundbetrieb. *Österreichische Zeitschrift für Elektrizitätswirtschaft*, 9(6), 277–281.

Varchmin, J. and Schubert, M. (1988) *Stromerzeugung und Elektrizitätswirtschaft. Aufstieg der Elektroindustrie – das Energiegesetz – Energieinsel Berlin.* Berlin: Museum für Verkehr und Technik.

Wan, C.T. (2014) Importing electricity could adversely affect cost and supply. *South China Morning Post.* Available at: http://www.scmp.com/comment/letters/article/1510121/china-southern-grid-electricity-imports-could-adversely-affect-cost (accessed 12 August 2015).

Waters, D. (1990) Hong Kong 'Hongs' with histories and British connections. *Journal of the Royal Asiatic Society Hong Kong Branch*, 30, 219–256.

Winje, D. (1994) Integration des West-Berliner Netzes in den deutschen Verbund. *Elektrizitätswirtschaft*, 93(13), 726–732.

Ziesing, H.J. (1985) Strukturelle und sektorale Entwicklung des Energieverbrauchs in Berlin (West). *Wochenbericht des Deutschen Instituts für Wirtschaftsforschung*, 52, 227–238.

5

THE AMENABLE CITY-REGION

The symbolic rise and the relative decline of Greater Manchester's low carbon commitments, 2006–17

Mike Hodson, Simon Marvin and Andy McMeekin

1 Introduction

This chapter contributes to an emerging but increasingly vibrant debate about the role of cities in infrastructure transitions to low carbon futures (Bridge et al., 2013; Hodson et al., 2013; Bulkeley et al., 2011). It starts from the view that networked infrastructures, that support the social and economic life of the city and that produce particular ecological consequences, can be shaped in various ways. In doing this, potentially different coalitions of social interest can claim to speak on behalf of the city. Consequently, this chapter's focus is on the intersection of two sets of issues. The first is the changing nature of relationships between city and networked infrastructures in a period of conflicting economic, ecological and political pressures and the search for a low carbon future. The second is the social interests, institutions and actors, who seek to shape such relationships and the ways in which they are organized to act in doing so.

Cities and regions are both integrated into wider networked infrastructure systems as key sites of consumption and implicated in hierarchies of multilevel governance. This emphasizes the importance of understanding relationships between scales and how and why they are coordinated, in tension or even disconnected. It involves examining the relationships between national priorities and plans and how these are interpreted and responded to at an urban level.

At this interface, our concern in this chapter is with visions and how they represent attempts to envisage and design low carbon urban futures. They do this through representing relationships between city, infrastructure and low carbon futures. We recognize the transient and politically produced nature of visions and, therefore, their mutability. Our analytical focus is a long-term understanding of how urban low carbon visions change over time and how this is shaped by social interests. In doing this, we illustrate not only how a low carbon future becomes envisaged and ascendant in a particular place but also how it becomes re-visioned and subject to relative decline.

The chapter does this through a case study of two different stages of attempts to undertake a low carbon transition in Greater Manchester. Greater Manchester has sought in recent decades to reposition itself from an industrial city to a post-industrial entrepreneurial city where new formal metropolitan governance arrangements that have been absent since 1986 have since the late 2000s been under development (Deas, 2014). This changing governance context has had implications for how plans to build low carbon capabilities have been framed and envisioned alongside a range of other, primarily economic, priorities. Power relationships between national government and 'local' social interests frequently favour the former. This leads to visions at the Greater Manchester level that represent a limited 'local' autonomy and that seek to represent the city-region as an amenable site for 'external' interests to come in and provide the resources and know-how to achieve low carbon transition.

The rest of the chapter is divided into six parts. First we review the literature on urban socio-technical transitions and locate the role of visions in shaping transition. Second, we use documentary analysis of national policy documents to identify how national priorities around energy and climate change created the context for shaping early low carbon visions in Greater Manchester. Third, we use documentary analysis, observational material and an interview programme with city-region-scale actors to set out how attempts to forge governance and institutions for a low carbon city-region created conditions to build a low carbon urban vision in Greater Manchester. Fourth, we illustrate long-term, shifting visions of a low carbon future in Greater Manchester. Exploring this in two phases (2006–11 and 2012–17) highlights a process starting with the symbolic rise and ending with the relative decline of a low carbon agenda in Greater Manchester. This involves a changing vision, with a shifting focus from energy to transport. This changing vision is guided by a focus that shifts from the perceived economic benefits of addressing climate change to a narrower version of state economic restructuring. Fifth, we summarize how these shifting visions were produced by 'local' amenability to UK national priorities. Finally, we finish by offering key conclusions. This sets out how the rise and fall of low carbon visions in Greater Manchester over time can be understood through relative structural weakness. Through its relationships with national government, in particular, Greater Manchester's low carbon visions were built on the basis of amenability to the interests of 'others'. This informed both the rise and the relative decline of low carbon in Greater Manchester.

2 Place-based low carbon transitions and the role of visions

Transitions analyses, both historical and prospective, have been put to work across a range of substantive technological, systemic and sectoral areas (Geels, 2004; Voß et al. 2006; Smith et al., 2010). In particular, analysing the extent to which transitions can be undertaken at the level of the city is important because of the complex pressures and tensions for urban economic growth, national targets for reducing subnational carbon emissions, and developing effective responses to the threats of climate change (Hodson and Marvin, 2010a). Urban authorities and wider coalitions, in particular in world cities, have sought to rescale energy systems and reconfigure transport systems at the level of the city, to try and assert control over the organization and functioning of energy and transport systems in order to build greater security for cities and the social interests who benefit from urban growth. The difficulty of this, in a UK context, is that the promotion of this competitive form of city-based economic growth takes place alongside national government prioritizing of particular cities.

There are important issues raised by the ways in which national governments view the role of cities and communities in undertaking transitions. Transitions in energy and transport systems, for example, at the city scale may look very different. In the case of energy this could mean rescaling a largely regionally and nationally organized energy regime, while in the case of transport, this could be reconfiguration of a local transport system that is nested within a wider national transport system. This may require the constitution of city-scale energy or transport regimes with variable levels of discretion afforded by national government. In this sense, when we talk about an urban low carbon transition, we are referring to a rescaling and/or reconfiguration of the city's energy and transport regimes. For urban stakeholders, from local authorities to business and communities, this requires the development of capacity to act in undertaking such a transition.

A number of contributions have highlighted that locating the role of the city – theoretically, conceptually and empirically – in low carbon transitions is extremely difficult (Hodson and Marvin, 2010a; Späth and Rohracher, 2010; Bulkeley et al., 2011; Coutard and Rutherford, 2010). A leading approach to addressing transitions, the multilevel perspective (MLP), says little explicitly about cities, who, what and where the city is, and their roles within transitions. In the MLP the spatial context for transition was often de facto national transition. Spatial issues were pregiven, underplayed or ignored. However, in recent years, geographers have fruitfully engaged with these debates (Bulkeley et al, 2011; Truffer and Coenen, 2012; Bridge et al, 2013; Hodson and Marvin, 2010a, 2010b; Raven et al., 2012; Murphy, 2015; Longhurst, 2015).

This engagement has produced attempts to bring together in various ways transitions in 'what' (systems) and 'where' (various spatial contexts) over the last decade. How urban transitions are accomplished (Bulkeley et al., 2011) is underpinned by the idea that scales and spaces of transitions are actively and relationally produced (Raven et al., 2012). This requires understanding institutional configurations shaping socio-technical processes and particular territorial spaces.

A key focus of this wider debate has been about the constitution of space and scale in transitions and the role of cities in transition processes (Hodson and Marvin, 2010a, 2010b). Of particular importance are the ways in which institutional and social interests are organized through governance processes that seek to shape urban transitions (Hodson et al., 2013). This involves not just territorially specific institutions and interests but also national and international institutions and networks of interests (Acuto, 2016; Bulkeley and Betsill, 2003).

Yet the framing of low carbon transitions can be manifold. The concept of 'vision' in the MLP is important as "the articulation of visions and expectations [functions] to provide an orientation towards the future and give direction to learning processes" (Geels, 2005, p. 366). Given the range of new and vested interests, it is important to recognize the struggles and negotiations that inform the production of visions. Visions, in this understanding, may be recast as symbolic representations of future relationships between city and regime. They are produced through relational struggles to define and categorize.

Low carbon urban transitions, in this view, can be understood through the visions that are constituted and the type of transition that is deemed to be required. In producing visions, the issue of which interests shape them is a key concern. In this regard, relationships between national-state and territorial interests are an important focus.

3 National low carbon policy priorities and the role of place

Understanding recent national low carbon/place policy priorities in the UK is far from straightforward. To take the area of energy policy, for example; until the Department of Energy and Climate Change (DECC) was created in 2008 there had not been a solely designated Department of Energy in the UK since 1992. This meant that historically, government priorities around energy were formulated in a multiplicity of departments, which had a range of issues as their core brief – trade and industry, environment, food and rural affairs, etc. The consequence of this is that, at the time of the development of the UK low carbon agenda from 2008, UK priorities around energy needed to be pieced together from a variety of different departmental positions in which energy priorities were mediated through four dominant strategies (see Table 5.1).

Setting parameters for urban responses through targets, plans and city-regions

The significance of the 2008 UK Climate Change Act was in its positioning of the UK as the first country in the world to have a legally binding framework for cutting carbon emissions. It does that through setting legally binding targets, creating powers to address those targets and providing the institutional framework to underpin the achievement of these targets (DEFRA, 2008). Among the key priorities in the Act is the setting of legally binding greenhouse gas emissions reduction targets of at least 80% by 2050 with an interim reduction in emissions of at least 34%, from a 1990 baseline, by 2020. This is to be achieved through five-year carbon budgeting systems. These developments create new pressures relating to climate change and carbon regulation (While, 2008). In addition to statutory carbon reduction targets cascaded down from international agreements (Bulkeley and Betsill, 2003), those developed by the national government place renewed

TABLE 5.1 UK energy priorities and their relationship to territory.

Plan	Priorities	Promoter	Spatial concept	Exemplification
Climate Change Act; Energy Act 2008	Binding, long-term statutory greenhouse gas emissions reductions – 80% 2050; 34% 2020 – carbon targeting and budgeting	DECC/DEFRA	Potential for budgeting, targets and the cascading of these down through various territorial tiers	Place-based low carbon budgets
UK Low Carbon Transition Plan (from July 2009)	Long-term transition plan to a UK low carbon future	DECC	Multiple views of places both implicitly and explicitly	Community pilots; Competitions for towns and cities
Low Carbon Industrial Strategy (from July 2009)	Low carbon industrial interventionism	BIS/DECC	Low carbon economic areas	North East England; South West England; M4 Corridor; Greater Manchester
Statutory City Region Pilots (from April 2009)	Design & piloting of city-regional governance structures for sustained economic growth	Treasury/DCLG	City regions	Manchester; Leeds

Note: BIS, Department for Business, Innovation and Skills; DCLG, Department for Communities and Local Government; DECC, Department of Energy and Climate Change; DEFRA, Department for Environment, Food and Rural Affairs.

emphasis on subnational territorial units, and will then place a premium on the ability of states and territories to better manage energy consumption and accelerate the development of low carbon energy transitions.

In July 2009, the UK government published its Low Carbon Transition Plan (LCTP) which detailed broadly how the UK would meet the 2020 and 2050 emissions reduction commitments set out in the Climate Change Act. LCTP was not only a transition route map to 2020 for the UK but also operated in prioritizing the carbon savings expected across different communities and cities. The Low Carbon Industrial Strategy was launched jointly by the Department for Business, Innovation and Skills and DECC in July 2009. Its aim was to position British businesses to secure the economic and job creation opportunities of a low carbon transition and, in doing so, to minimize the economic costs of inaction. It provided a more strategic approach to the development of low carbon economic activity and technologies across the regions of the UK, particularly through designating low carbon economic areas.

Through this industrial interventionism, the national government sought to shape subnational low carbon activities. In 2008 the UK government sought to move beyond this and to create the governance structures through which subnational areas could create 'their own' capacity to respond to the carbon agenda. This was enacted via the UK government's support for the creation of two city-regions (Leeds and Greater Manchester) in 2009 and, in doing so, the development of new metropolitan governance structures. The broad parameters within which the city-regions were to operate was as statutory forms of subregional cooperation between local authorities with the aim of them being significant contributors to sustainable forms of economic growth. Low carbon economic activities were also worked into these proposals as the city-regions took a more active role in shaping low carbon transition in their own contexts.

From targets and plans to practice: new industrial interventionism or redesigning subnational governance?

The priorities of the Climate Change Act and its emphasis on emissions reduction were, at the time, broadly supported across UK political parties. There were, similarly, few dissenting political voices in relation to the LCTP. The principal political tension was in the process of how the strategic priorities were to be achieved – what were to be the mediating frameworks and institutions and what economic, social and knowledge resources were to be allocated to them?

The tension was between, first, new forms of state industrial interventionism in regions, city-regions and pan-regions and, second, in constructing new forms of national–subnational governance fixes that saw the state less in the direct role of industrial intervener and more as a 'facilitator' of market-based activity for city-regions. This struggle was inherent within the Labour government, which governed until 2010, and cut across its different strategies. The subsequent coalition government, from May 2010, embarked on a comprehensive redesign of subnational governance. It actively sought to abolish and redesign institutional mediators between its central departments and places with the aim of creating the conditions to compete for limited resources and create private and entrepreneurial low carbon responses. In short, the existing dominant mediators of national–subnational relations – Regional Development Agencies – were abolished and replaced by Local Enterprise Partnerships. At the same time a much less well resourced Regional Investment Fund – circa 1 billion GBP (Great Britain pounds) – sought to intensify competition between places for national resources and support.

The national policies and priorities that we have reviewed encompass a range of economic, environmental, technological and territorial issues. It is in this context that territorial visions of low carbon urban futures were developed. The particular ways in which different priorities coalesce within the context of a city-region to produce a vision are unclear. Building understanding of this requires addressing how these different national priorities create the conditions for city-regional low carbon visions. It is with this as our focus, that we now address the development of low carbon visions in Greater Manchester over time.

4 The symbolic rise and relative decline of low carbon in Greater Manchester

The national priorities and strategies outlined above set the parameters within which understanding visions at the city scale need to be framed. In this section, we address the development of low carbon visions in Greater Manchester. To do this, we focus on a time period from the early days of development of a low carbon agenda in Greater Manchester (2006) through to its relative decline (2017). We set out two visions of a low carbon Greater Manchester. The first of these (2006–11) highlights the symbolic rise of a low carbon energy agenda in Greater Manchester. This was promoted as a means through which low carbon economic growth could be promoted and pursued. This vision was strongly conditioned by national priorities. As national priorities began to shift with a change of UK government, a second vision (2012–17) developed. This was less focused on carbon reduction and much more directly concerned with economic growth. The vision promoted state-spatial and urban governance restructuring as a means through which economic growth could be achieved. Transitions in transport – rather than energy – infrastructure were central to the vision. The shifting visions highlight the sensitivity of Greater Manchester institutions to changing national priorities. In this respect, 'local' priorities and visions were couched in a wider strategy of making the city-region amenable to the priorities of national government.

2006–11: the symbolic rise of low carbon

'Vision': climate change as a low carbon economic opportunity for Greater Manchester and meeting national targets

The Association of Greater Manchester Authorities (AGMA), from 1986 onwards, worked to coordinate the actions of the ten constituent Greater Manchester local authorities on cross-boundary strategic areas – such as transport and waste – that were seen as most effectively organized at a metropolitan scale. In January 2008 the Executive of AGMA agreed in principle to support the establishment of a Climate Change Agency (CCA) across the city-region. The development of the CCA was the product of a confluence of activities that sought to coordinate local actions and 'sustainable' economic strategy at a city-regional scale. A significant step in this was the 2008 Mini-Stern for Manchester commissioned by AGMA and undertaken by the consultants Deloitte (Deloitte, 2008). It aimed to produce a rescaled version of the Stern Report that was written for the UK Treasury in 2006, by the economist Nicholas Stern, reviewing the economics of climate change (Stern, 2006).

   The Mini-Stern sought to calculate the cost of climate change for the Manchester city-region and possible strategic responses in the short- to medium-term particularly given international pressures and national legislation on climate change. The report was framed relatively narrowly in that it examined the economic costs of inaction on climate change and the economic opportunities of early action. On the basis of analysis conducted using a methodology developed by Deloitte, it calculated that inaction on climate change would potentially cost Greater Manchester 21 billion GBP over a 12-year period and 72 billion GBP at the North West regional level. It prescribed a necessity for action at the Greater Manchester level that prioritized distinctiveness, first-mover advantages, technology-led responses, business and investment, business support, and attempts to attract inward investment. The focus on energy in the review was oriented towards reshaping city-regional energy flows in line with meeting renewable energy and other national targets.

Consequently, it argued for a strategic approach to energy production, consumption and the reconfiguration of Greater Manchester's energy system through energy efficiencies and new technologies to reduce economic cost and meet carbon reduction targets. In this respect, plans to achieve national targets, while also meeting the city-region's energy 'needs' were linked to achieving greater control over energy production and consumption relationships through coordination of urban regeneration activities and latent innovation capacity in Greater Manchester. Mini-Stern also raised the importance of eco-innovation and creating new markets for services and technologies. In doing this, the role of the public sector was promoted as an exemplar in terms of procurement strategies and its own estate management. This was also part of an attempt to create a culture that was conducive to business investment, inward investment and the availability of support and advice. In doing this there was also an emphasis on developing and building skills and capacity – with a role for higher education institutions – and the alignment of policies. The logic of the report appeared to be in extending the economic competition between places into the sphere of climate change and for Greater Manchester to seek to exploit a first mover advantage and distinctiveness in eco-economic competition. To address many of these issues and concerns the Climate Change Agency was set up in parallel with two other energy intermediaries.

From around 2008 there was a meshing, by Greater Manchester elites, of a strategy of promoting an entrepreneurial economic agenda with efforts to build a low carbon agenda in Greater Manchester. By 2011 new governance arrangements for Greater Manchester incorporated the wider environmental agenda in an Environment Commission, which subsequently became rebranded as the Low Carbon Hub. At this time, the Greater Manchester Climate Change Strategy was produced and set targets for a 48% reduction in CO_2 emissions by 2020 (AGMA, 2011). The strategy's narrative clearly reflected the Mini-Stern approach of incorporating economic and environmental logics, setting out the 'boost' to the economy and jobs from the transition to a low carbon economy and addressing CO_2 emissions. The implementation of the strategy covered various infrastructural interventions around energy, the built environment, transport and so on. Though there is something of a lag in being able to access up-to-date data, recent data suggest that by 2013 there had been a 27.5% reduction from 1990 in Greater Manchester's CO_2 emissions. This was slightly behind target (GMCA/AGMA, 2016).

Greater Manchester's decision-makers have constrained discretion to meet these targets. This is the case in relation to governance processes where, even in a context of ongoing reorganization of national–city-regional relations from 2011, the vast majority of public funding for Greater Manchester remains in the control of the UK state, and where the governance of infrastructure systems does not correspond with the territorial boundaries of Greater Manchester.

2012–17: the relative decline of low carbon

'Vision': the state-spatial economic restructuring of Greater Manchester

The constrained discretion, set out above, is not a static situation. From 2012, a City Deal (GMCA, 2012) between the UK national government and Greater Manchester was intended to be a tailored agreement to devolve funding and decision-making on the basis of contributing to the promotion of local economic growth. This produced a vision of the economic restructuring of Greater Manchester. Its focus was on accelerated economic growth, promoted in ways where carbon issues appeared to become relatively less prioritized. Transport infrastructure became a key part of this vision of economic restructuring.

By November 2014, the UK Chancellor of the Exchequer, George Osborne, and the leaders of Greater Manchester local authorities were together in Manchester Town Hall presenting a package of devolved measures (including powers related to transport, planning, housing, skills and economic development). The deal was reckoned to be worth more than 1 billion GBP; still a fraction of the 22 billion GBP public spending attributed to Greater Manchester; it was followed in February of 2015 by an agreement that around 6 billion GBP of the national health budget would be devolved to Greater Manchester.

The deal, characterized as 'Devo Manc', has produced a slew of positive symbolism but there remains much uncertainty as to what its tangible, long-term implications will be. Its dominant representation is of a restructuring that sees a future Greater Manchester at the centre of a Northern Powerhouse of connected northern English metropolitan areas. The stated aim is to address the UK's economic imbalance towards London and the South East of England and to create a second growth pole able to compete with global economic powerhouses. It promotes agglomeration, which implies prioritizing particular areas and sectors of 'strength' rather than the whole of Greater Manchester. The implication is that Devo Manc will produce a selective rather than a comprehensive reconfiguration of Greater Manchester's infrastructure predicated principally on enhanced economic growth. But Devo Manc is also a means of connecting (parts of) Greater Manchester to wider circuits and flows. What is clear is that, symbolically at least, this involves promotion of both agglomeration-based economic growth and the creation of new, selective connections to wider global circuits, flows and operational landscapes. In short, an acceleration of growth is prioritized. Yet, there is limited concrete sense of how the carbon emissions associated with this extra growth will be addressed. In this respect, the growth agenda is enhanced while the low carbon agenda is relatively marginalized.

The issue here is in what ways are plans to restructure Greater Manchester made material? In 2016, for the first time since 1981, a metropolitan spatial plan (the Greater Manchester Spatial Framework (GMSF)) was put out for consultation by the Greater Manchester Combined Authority (GMCA, 2016). The GMSF sets out strategic orientation in relation to housing, land use, employment and associated infrastructure for the next twenty years. In this sense, it provides a systematic, long-term attempt to reconfigure the city-region – creating new path dependencies and structuring urban space through new material infrastructures, employment, leisure activities and ways of living. Its primary aim is to address "how Greater Manchester is planning to meet levels of growth well above baseline forecasts" (GMCA, 2016, p. 6). It also sets out site allocations to achieve these strategic aims.

The focus of the GMSF is on productivity and the acceleration of growth in Greater Manchester, underpinning a "vision for the future [that] is of an even more successful Greater Manchester, which can compete on the global stage to attract investment, businesses, workers and tourists" (GMCA, 2016, p. 11). The acceleration of growth envisaged for Greater Manchester takes place in a context of relatively low economic growth in Western national states since the financial crisis of 2007/ 8. The assumption is that through purposive interventions, Greater Manchester can buck this trend and produce growth 'above baseline conditions' of 5 billion GBP by 2035, underpinned by population growth of 294,800, creating an additional 199,700 jobs and requiring 227,200 net new homes. Symbolic investment in transport infrastructure, specifically high-speed rail links between London and Manchester and across the north of England and the growth of the airport, is trumpeted. The approach to accelerating growth is through parcelling up spaces of multiple kinds of activities (housing, retail, warehousing) and combinations of development opportunities to attract inward investment.

Reconfiguring Greater Manchester is not a spatially even process. Representationally, the GMSF framework suggests that "[a]ll parts of Greater Manchester will make a contribution towards growth and prosperity, but there is a small number of locations that will be strategically significant in terms of their economic importance and role in meeting future development needs" (GMCA, 2016, p. 15). All places are important, just that some are more important than others! Getting beyond this contradiction of being both comprehensive *and* selective, the GMSF makes the claim, more in keeping with its content, that "there will continue to be a very strong focus on the core of the conurbation" (GMCA, 2016, p. 12). This focus on office, residential and retail development prioritizes Manchester city centre, and, additionally, the now regenerated old docklands area of Salford Quays, and the airport. There is also prioritization of numerous 'gateways' and 'corridors' that attempt to reinforce existing strengths and transport interconnections through intensifying warehousing and logistics capacity. The reformatting of the remainder of the city-region remains at best underdeveloped and worst unspecified.

5 Producing visions in the amenable city: squeezing low carbon

Two different visions have promoted the symbolic rise and the relative decline of the low carbon agenda in Greater Manchester. What is common to both visions is that they have 'shadowed' and been strongly shaped by UK national priorities. In both these visions, UK state restructuring of Greater Manchester's institutional and governance architectures has been strategically connected to the contextually selective interpretation and repackaging of urban infrastructure. This repackaging is communicated through visions that first prioritized a central role for selective energy infrastructures and subsequently transport infrastructures. Rather than pursuing a broad and inclusive form of infrastructural reconfiguration, preparatory governance work appears to aim to make Greater Manchester amenable to uneven infrastructuralized change.

There remains a low carbon strategy, targets and an implementation plan but the effect of accelerating growth is a relative squeezing of low carbon concerns. On carbon emissions, Greater Manchester's approach to low carbon, developed initially in 2008, became aligned to some extent with Greater Manchester's growth strategy; the approach to carbon became about how a focus on it could enhance green growth. There were and are commitments to address carbon emissions through combinations of market opportunities (e.g. retrofitting of the built environment), through large-scale infrastructure investment (e.g. transport infrastructure), through rescaling and refocusing resource flows (e.g. decentralized heating) "as well as attempting to lock-in larger quantities of carbon through habitat restoration and expansion" (GMCA, 2016, p. 81). Greater Manchester leaders are sensitive to the competitive metrics and peer pressure that operate in relation to carbon. The entrepreneurial urbanism, dominant since the 1980s, is manifest in Greater Manchester's approach to carbon emissions reduction.

The GMSF claims that there will be a 60% reduction in carbon emissions by 2035 from 1990 levels. Yet, work undertaken for GMCA suggests that "with the right combination of drivers and action" a 59% reduction could be achieved. But, even "[t]his will be very challenging to achieve" (GMCA, 2016, p. 81). The issue is not that carbon emissions are ignored in the framework. The priority is accelerating economic growth. This reconfiguration of the city-region is governed through public governing institutions creating the conditions for private investment and either filling in where private investment can't see a return or leaving places to muddle through. The plan is to reconfigure the city-region by creating opportunities for developers to remake it where public–private governance prioritizes selective (premium and logistical) spaces. In this approach, although there remains a focus on carbon, there is already a struggle to meet carbon targets. Even optimistic scenarios fall some way short. The GMSF framework promotes growth significantly above baseline. This suggests that carbon emissions reductions will be even more difficult to achieve and lowering carbon, as an agenda, is becoming more squeezed.

The significance of these visions is that they require discursive preparation and justification by urban governance interests. In Greater Manchester this often involves coalitions of metropolitan and national political elites mobilizing a narrow agenda of how to 'entrepreneurially' use infrastructure to reposition Greater Manchester. The two visions highlight the different ways in which this was addressed over time. The consequence of this is that wider sustainability concerns – that were apparent in the 1990s and which, in the 2000s in a UK context, promoted a low carbon agenda – are being squeezed (Hodson and Marvin, 2014) and accelerated economic growth is prioritized.

6 Conclusions

In this chapter we have used research conducted in Greater Manchester to understand visions of urban low carbon futures. We explored this through focusing on two phases (2006–11 and 2012–17) of low carbon development in Greater Manchester that started with the symbolic rise and ended with the relative demise of a low carbon agenda. This shift can be understood through relative structural weakness and the dependency on national government of Greater Manchester's governing institutions. Through its relationships with national government, Greater Manchester's low carbon visions were built on the basis of amenability to the interests of 'others'.

The contribution of the chapter to understanding urban low carbon transitions is threefold. Firstly, we have sought to develop a better understanding of the dynamics of the relationships between national and city-regional priorities in producing visions of low carbon urban futures. In particular we wanted to develop a more sophisticated appreciation of the ways in which urban low carbon transitions are constituted as metropolitan-level responses to national priorities. Secondly, our interest was in what shape these responses take, whether they are piecemeal or more strategic, and who was involved and excluded. We have shown that national priorities have powerfully shaped the city-regional responses in Greater Manchester. Thirdly, the focus on visions is not just about technological artefacts at a city scale but also the social interests, institutions and actors, who seek to shape a city-region through infrastructural reconfiguration. In Greater Manchester, the additional complication was that city-regional governance structures were themselves being remade with a central role for, initially, climate change and energy governance and then subsequently a more boldly economic growth strategy and the transport infrastructure that would support it. In doing this we were keen to demonstrate, through the concept of visions, the messy politics of these responses, the dominance of national and economic priorities that were being advocated and the narrowness of the urban transition that might be produced if these visions were accomplished.

Finally, in thinking about future research agendas and visions of low carbon urban futures three issues would be worthy of further research. Firstly, it is important to ask in what different ways the current confluence of economic and ecological crises and pressures are being used to influence existing commitments and future development of low carbon urban transitions. Are they being used to defend, challenge or weaken the status quo? Secondly, to what extent do these crises and pressures contribute to or constrain the possibilities for different notions of low carbon urban transitions. Is there still potential to develop visions of low carbon futures as a participatory process? How are these to be organized and with what consequences? Finally, it would be valuable to explore whether low carbon urban visions will necessarily be dominated by national priorities. What are the possibilities for other wider city-regional priorities and grass-roots and community initiatives to inform such transitions?

Acknowledgements

Parts of this chapter draw on material from: Hodson, M., and Marvin, S. (2012) 'Mediating low-carbon urban transitions? Forms of organisation, knowledge and action', *European Planning Studies*, 20(3), 421–439. Mike Hodson and Andy McMeekin are grateful for the support of the Alliance Manchester Business School Strategic Research Investment Fund for supporting research in this chapter. Our thanks to Jon Silver for helpful conversations about the shape of this chapter and also to the editors for suggesting improvements to the chapter.

References

Acuto, M. (2016) Retrofitting global environmental politics: Networking and climate action in the C40. In: Hodson, M. and Marvin, S. (ed.) *Retrofitting Cities: Priorities, Governance and Experimentation*. London: Routledge, pp. 107–118.

AGMA (Association of Greater Manchester Authorities) (2009a) *Manchester Statutory City Region Agreement*. Manchester: AGMA.

AGMA (Association of Greater Manchester Authorities) (2009b) *Prosperity for All: The Greater Manchester Strategy*. August. Manchester: AGMA.

AGMA (Association of Greater Manchester Authorities) (2011) *Implementation Plan 2012–2015, Greater Manchester Climate Change Strategy*. Manchester: AGMA.

AGMA Executive Board (2009) *Statutory City Regions and Greater Manchester Strategy*. Manchester: AGMA.

AGMA and Manchester Enterprises (2008) *The Manchester Multi-area Agreement: Our City Region's Proposal to Government*. Manchester: AGMA.

BERR (Department for Business, Enterprise and Regulatory Reform) (2007) *Energy White Paper: Meeting the Energy Challenge*. London: HMSO.

BIS/DECC (Department for Business, Innovation and Skills /Department of Energy and Climate Change) (2009) *Low Carbon Innovation Strategy: A Vision*. London: HM Government.

Bridge, G., Bouzarovski, S., Bradshaw, M. and Eyre, N. (2013) Geographies of energy transition: space, place and the low-carbon economy. *Energy Policy*, 53, 331–340.

Bulkeley, H. and Betsill, M. (2003) *Cities and Climate Change: Urban Sustainability and Global Environmental Governance*. New York: Routledge.

Bulkeley, H., Castán Broto, V., Hodson, M. and Marvin, S. (eds.) (2011) *Cities and Low Carbon Transitions*. London: Routledge.

Committee on Climate Change (2008) *Building a Low-Carbon Economy – The UK's Contribution to Tackling Climate Change: The First Report of the Committee on Climate Change*. London: Committee on Climate Change.

Coutard, O. and Rutherford, J. (2010) Energy transition and city-region planning: understanding the spatial politics of systemic change. *Technology Analysis and Strategic Management*, 22(6), 711–727.

DCLG (Department for Communities and Local Government) (2007) *The Review of Sub-national Economic Development and Regeneration Sub-national Review*. London: HM Government.

DCLG (Department for Communities and Local Government) (2008) *Explanatory Note on Taking Forward Proposals on City-Regions in PBR 2008*. London: DCLG.

Deas, I. (2014) The search for territorial fixes in subnational governance: city-regions and the disputed emergence of post-political consensus in Manchester, England. *Urban Studies*, 51(11), 2285–2314.

DECC (Department of Energy and Climate Change) (2009a) *The UK Low Carbon Transition Plan*. London: HM Government.

DECC (Department of Energy and Climate Change) (2009b) *The UK Renewable Energy Strategy*. London: HM Government.

DEFRA (Department for Environment, Food and Rural Affairs) (2008) *Climate Change Act 2008*. London: HMSO. http://www.legislation.gov.uk/ukpga/2008/27/pdfs/ukpga_20080027_en.pdf (accessed 20 July 2017).

Deloitte (2008) *'Mini Stern' for Manchester: Assessing the Economic Impact of EU and UK Climate Change Legislation on Manchester City Region and the North West*. London: Deloitte. Available at: http://media.ontheplatform.org.uk/sites/default/files/UK_GPS_MiniStern.pdf (accessed 20 July 2017).

Geels, F. (2002) Technological transitions as evolutionary reconfiguration processes: a multi-level perspective and a case study. *Research Policy*, 31(8), 1257–1274.

Geels, F. (2004) From sectoral systems of innovation to socio-technical systems: insights about dynamics and change from sociology and institutional theory. *Research Policy*, 33(6–7), 897–920.

GeelsF. (2005) Co-evolution of technology and society: the transition in water supply and personal hygiene in the Netherlands (1850–1930) – a case study in multi-level perspective. *Technology in Society*, 27(3), 363–397.

Geels, F. (2011) The role of cities in technological transitions: analytical clarifications and historical examples. In: Bulkeley, H., Castán-Broto, V., Hodson, M. and Marvin, S., (eds.) *Cities and Low Carbon Transitions*. London: Routledge, pp. 13–27.

GMCA (Greater Manchester Combined Authority) (2012) *Greater Manchester City Deal*. Manchester: GMCA.

GMCA (Greater Manchester Combined Authority) (2016) *Draft Greater Manchester Spatial Framework*. Manchester: GMCA.

GMCA/AGMA (Greater Manchester Combined Authority/Association of Greater Manchester Authorities) (2016) *Joint Greater Manchester Combined Authority and AGMA Executive Board, Greater Manchester Strategy Annual Performance Report, Report of: Tony Lloyd, GM Interim Mayor of Greater Manchester and Sir Howard Bernstein, Head of Paid Service, 29 July*. Manchester: GMCA.

GMEC (Greater Manchester Environment Commission) (2009a) *Delivering An Energy Group And Plan For Greater Manchester: An Update On Progress*. Report to Environment Commission. 10 September. Manchester: GMEC.

GMEC (Greater Manchester Environment Commission) (2009b) *Infrastructure Cross Commission Working Report to Environment Commission*. 10 September. Manchester: GMEC.

GMEC (Greater Manchester Environment Commission) (2009c) *Minutes Environment Commission*. 10 September. Manchester: GMEC.

GMEC (Greater Manchester Environment Commission) (2009d) *Report to Environment Commission on Environment Commission Away Day – 8 June 2009*. 14 May. Manchester: GMEC.

GMEC (Greater Manchester Environment Commission) (2009e) *Report to Environment Commission on Climate Change Agency*. 14 May. Manchester: GMEC.

GMEC (Greater Manchester Environment Commission) (2009f) *Report to Environment Commission on Energy Planning in Greater Manchester*. 29 May. Manchester: GMEC.

GMEC (Greater Manchester Environment Commission) (2009g) *Report to Environment Commission on The Transition to a Low Carbon Economy*. 16 July. Manchester: GMEC.

GMEC (Greater Manchester Environment Commission) (2009h) *Report to Environment Commission on Environment Commission Work Programme: Commissioner Roles*. 2 November. Manchester: GMEC.

GMEC (Greater Manchester Environment Commission) (2009i) *Report to Environment Commission on Low Carbon Economic Area and CCA Update*. 2 November. Manchester: GMEC.

GMEC (Greater Manchester Environment Commission) (2010a) *Report to Environment Commission on Low Carbon Economic Area*. 20 January. Manchester: GMEC.

GMEC (Greater Manchester Environment Commission) (2010b) *Report to Environment Commission on Low Carbon Economic Area*. 18 March. Manchester: GMEC.

GMEC (Greater Manchester Environment Commission) (2010c) *Work Programme Implementation Plan*. 20 January. Manchester: GMEC.

GMPHC (Greater Manchester Planning and Housing Commission) (2010) *Report to Planning and Housing Commission on Next Steps for the AGMA Decentralised/Low Carbon Energy Study*. 17 March. Manchester: GMPHC.

Hodson, M. and Marvin, S. (2010a) *World Cities and Climate Change*. Maidenhead: McGraw-Hill.

Hodson, M. and Marvin, S. (2010b) Can cities shape socio-technical transitions and how would we know if they were? *Research Policy*, 39(4), 477–485.

Hodson, M. and Marvin, S. (2014) Introduction. In: Hodson, M. and Marvin, S. (eds.) *After Sustainable Cities?* London: Routledge, pp. 1–9.

Hodson, M., Marvin, S. and Bulkeley, H. (2013) The intermediary organisation of low carbon cities: a comparative analysis of transitions in Greater London and Greater Manchester. *Urban Studies*, 50(7), 1403–1422.

Hodson, M., Marvin, S., Bulkeley, H. and Castán Broto, V. (2011) Conclusion. In: Bulkeley, H., Castán-Broto, V., Hodson, M. and Marvin, S. (eds.) *Cities and Low Carbon Transitions*. London: Routledge, pp. 198–202.

HM Treasury (2009) *Budget 2009: Building Britain's Future*. London: Stationary Office.

HM Treasury (2008) *Pre-Budget Report 2008: facing global challenges supporting people through difficult times*. London: Stationary Office.

Longhurst, N. (2015) Towards an 'alternative' geography of innovation: alternative milieu, socio-cognitive protection and sustainability experimentation. *Environmental Innovation and Societal Transitions*, 17, 183–198.

McKillop, T., Coyle, D., Glaeser, E., Kestenbaum, J. and O'Neill, J. (2009) *Manchester Independent Economic Review*. Manchester: Centre for Local Economic Strategies.

Murphy, J.T. (2015) Human geography and socio-technical transition studies: Promising intersections. *Environmental Innovation and Societal Transitions*, 17, 73–91.

Peck, J. (2012) Austerity urbanism: American cities under extreme economy. *City*, 16(6), 626–655.

Raven, R.P.J.M., Schot, J. and Berkhout, F. (2012) Space and scale in socio-technical transitions. *Environmental Innovation and Societal Transitions*, 4, 63–78.

Smith, A., Voβ, J.P. and Grin, J. (2010) Innovation studies and sustainability transitions: the allure of the multi-level perspective, and its challenges. *Research Policy*, 39(4), 435–448.

Späth, P. and Rohracher, H. (2010) 'Energy regions': the transformative power of regional discourses on socio-technical futures. *Research Policy*, 39(4), 449–458.

Stern, N. (2006) *Stern Review on the Economic of Climate Change*. London: HM Treasury.

Transport for the North (2016) *The Northern Transport Strategy: Spring 2016 Report*. Available at: https://www.gov.uk/government/uploads/system/uploads/attachment_data/file/505705/northern-transport-strategy-spring-2016.pdf (accessed 20 July 2017).

Truffer, B. and Coenen, L. (2012) Environmental innovation and sustainability transitions in regional studies. *Regional Studies*, 46(1), 1–21.

Voß, J.-P., Bauknecht, D. and Kemp, R. (eds.) (2006) Sustainability and reflexive governance: introduction. In: Voß, J.-P., Bauknecht, D. and Kemp, R. (eds.) *Reflexive Governance for Sustainable Development*. Cheltenham: Edward Elgar, pp. 3–31.

While, A. (2008) Climate change and planning: carbon control and spatial regulation. *Town Planning Review*, 79(1), vii–xiii.

6

WHAT IS 'CARBON NEUTRAL'?

Planning urban deep decarbonization in North America

Laura Tozer

1 Introduction

Greenhouse gas (GHG) emissions need to be reduced to net zero by mid-century in order to retain the possibility of limiting global warming to 1.5°C (Rogelj et al., 2015). The vast scale of GHG emissions reduction required to achieve this target has significant implications for how we envision cities and urban life. Though this target requires net zero GHG emissions globally, it has been suggested that different urban areas would use different low carbon transformation strategies (IPCC, 2015). Therefore, it is not clear what decarbonized urban systems might be like. This chapter begins to answer that question by looking at various North American cities that have started to plan their decarbonized urban futures. In particular, through an analysis of official documents, it examines how the idea of 'carbon neutrality' within the building and energy sector is represented in nine North American local authorities striving for "deep decarbonization" (CNCA, 2015). In considering the development of novel ways of thinking about the low carbon–development interface, and the ways in which carbon is problematized in relation to several urban contexts (Luque-Ayala et al., Chapter 2, this volume), this chapter focuses on the design of low carbon urbanism.

The chapter first examines how carbon is problematized in cities to consider different ways of thinking about low carbon. Urban decarbonization is not restricted to setting targets and accounting for GHG emissions. Instead, it is a multifaceted process embedded in urban development. This process can be dominated by particular and different carbon problematization logics, and, as argued in this chapter, this influences the nature and extent of the low carbon transition. For example, a local energy utility company could frame the carbon problem as a matter of fossil fuel use, a position that would drive extensive overhauls of urban energy systems. Or perhaps a municipality could frame the problem as a matter of achieving carbon neutrality as soon as possible, driving small changes in the urban energy system combined with the acquisition of carbon offsets. Three vignettes on Vancouver, San Francisco and New York City included here illustrate differences in carbon problematization logics and the implications for urban development pathways. Out of this analysis, the chapter argues that variations in the interpretation of carbon neutrality create different problems requiring different solutions, which ultimately affect the socio-technical nature of actual low carbon urbanism. In other words, variations in the problematization of carbon can lead to different urban socio-technical paths. The examination of various low carbon logics provides insight into how urban development might be reshaped; but what future carbon neutral cities are imagined to be is yet to be discussed.

In its second half, the chapter looks at what kinds of elements urban actors plan to use to disrupt and disentangle carbon from urban flows. This acknowledges that it is not just ways of thinking about carbon that influence the design of low carbon urbanism, but also the selection of materials, technologies and infrastructures. In the making of low carbon urbanism, actors seek to intervene in carbon flows through particular assemblages of elements. For example, an emphasis on district energy and biofuels would likely create a different kind of low carbon city than an emphasis on solar photovoltaics (solar PV) and building energy efficiency. With this in mind, the chapter uses textual network analysis to examine the socio-technical elements that a group of North American cities "committed to achieving aggressive long-term carbon reduction goals" (CNCA, n.d.) plan to mobilize in order to become low carbon in the buildings and energy sector. The analysis shows that particular socio-technical interventions are being favoured in how low carbon urbanism is imagined in North America. Therefore, I argue that the materiality and socio-political context of different urban geographies impacts the design of low carbon urbanism. When drawn together, the various sections of the chapter unpack the interplay between underlying logics and the materiality of urban geographies in the design of low carbon urbanism.

The chapter is based on an analysis of the climate governance frameworks of the seventeen founding members of the Carbon Neutral Cities Alliance (CNCA), a transnational network of cities working towards cutting GHG emissions by 80% by 2050. The study was carried out in fall 2015 using discourse analysis and textual network analysis methods. The urban climate governance documents analysed were written between 2009 and 2015, and encompass a variety of policy materials, from carbon neutral strategy documents and climate change policy documents to renewable energy strategies. Although the policy documents were municipal-led, they frequently involved a range of stakeholders invoking diverse authorities. CNCA was founded in 2014 by a group of local government members of the C40 Cities Climate Leadership Group. CNCA members sought a transnational, networked space to focus on the particular challenges of what they refer to as a 'deep decarbonization' – a target of at least an 80% reduction of GHG emissions by 2050 (CNCA, 2015). The CNCA network offers grants to facilitate deep decarbonization innovation and has provided learning opportunities to members through peer-to-peer mentoring and the development of technical resources on decarbonization topics (CNCA, 2015). Though CNCA member city governments have been advancing climate responses for many years, at the time of writing 'deep decarbonization' activities were still largely confined to the planning stages.

While transnational networks have been important in the evolution of urban responses to climate change (Bulkeley and Betsill, 2013), the political landscape of transnational municipal climate networks is uneven. Some studies have questioned the extensiveness of the impact of some networks (Kern and Bulkeley, 2009); others focus on the various logics that drive such networks, ranging from reinforcement of neo-liberal economic principles to more radical goals (Toly, 2008; Bulkeley and Betsill, 2013). Drawing on these and other studies, and focusing on the North American members of CNCA, the chapter contributes to an analysis of how urban carbon neutral built environments are being imagined and designed. The chapter is divided into four sections, as follows: the first section reviews the literature on decarbonization and low carbon transformation; the second section introduces the CNCA; the third section considers the different ways in which carbon is being problematized, and what this variation means for integrating low carbon logics into urban development; the fourth and final section unpacks the socio-technical elements that are planned to be mobilized to achieve low carbon urban built environments of the future in North America.

2 Towards transformation

Urban decarbonization will require wide-reaching transformations to the material infrastructures of the city, as well as to its political and institutional systems, to achieve new low carbon configurations (Bulkeley et al., 2011). Decarbonization is about transformation. It requires fundamental changes to society aimed at disrupting the complex entanglements of carbon across technological and institutional systems (Unruh, 2000). As a result, decarbonization is a political process with significant scope beyond the practice of identifying efficient policies and innovative technologies (Bernstein and Hoffmann, 2016). Although many local authorities around the world have set goals to reduce GHG emissions, in actual practice they have largely taken a piecemeal approach that does not adequately address the key drivers of climate change (Romero-Lankao, 2012; Bulkeley and Betsill, 2013; Betsill and Bulkeley, 2007). Urban responses to climate change – often referred to in the academic literature as 'urban climate experiments' – encompass a broad range of different practices, some of which are capable of fostering systemic change while others are limited in their possibility to challenge fundamental ideas about production and consumption (Castán Broto and Bulkeley, 2013). Unable to advance systemic approaches (e.g. embedding low carbon logics within urban planning), urban low carbon interventions typically focus on individual sectors (e.g. building insulation, bike lanes, efficient heating, etc.) (Reckien et al., 2013). These findings paint a picture of incremental change rather than radical transformation. Though the scale of the transformation is difficult to assess, such types of analytical logics provide valuable tools to consider the potential direction and impact of the transformation at play.

Urban transformation does not depend on climate policy alone, but on the underlying development pathways of communities (Burch et al., 2014). As Luque-Ayala et al. argue in Chapter 2, low carbon transitions are multiple, non-linear and rhizomatic. This multiple nature, alongside the social, political, technical and economic entanglements of the transition, suggests that rather than a single pathway multiple pathways for low carbon are likely to be pursued. Rather than viewing this diversity as a competition to pluck out best practices, it is more analytically constructive to consider how low carbon is being variously embedded in urban development (Luque-Ayala et al., Chapter 2, this volume). In particular, the logics underpinning change within urban development pathways can provide critical insight into this urban transformation.

Sustainability transformation can be driven by different logics. For example, Scoones et al. (2015) identify four broad but common frames for green transformation: technocentric, marketized, state-led, and citizen-led. All of these narratives might be present, lacking or overlapping in a particular case of transition, but what is important is that invoked narratives represent various pathways to transformation; by triggering different problems and solutions, such narratives support the development of different transformations with varied power relations (Scoones et al., 2015). Similarly, we can consider how different logics shape low carbon urbanism. Over the course of this analysis, however, it is important not to lose sight of the materiality of the city. Such materiality, determining the entanglement of carbon in urban flows, impacts the shape of transition. Therefore, a consideration of the transformative potential of low carbon urbanism needs to take into account not just narratives, but also the material implications of low carbon practices (Lovell et al., 2009).

3 The Carbon Neutral Cities Alliance

In 2014, representatives from seventeen local governments gathered in Copenhagen to start a network called the Carbon Neutral Cities Alliance. The network is made up of local governments that are, in their own words, adopting "the most aggressive GHG reduction targets undertaken by any cities across the globe" (CNCA, 2015, p. 2). Local governments often participate in transnational governance networks to support their efforts to reduce GHG emissions locally (Bulkeley et al., 2014), but CNCA is distinctive from other transnational municipal governance networks because its members aim to achieve greater emissions reductions than those embraced by members of other networks. As self-identified forerunners of urban "deep decarbonization" (CNCA, 2015), the founding members of CNCA are establishing what it means for a city to become carbon neutral. North American local authorities, the focus of this chapter, make up more than half of the founding membership of CNCA (Table 6.1).

Before considering exactly what carbon neutral means in the context of CNCA, it is useful to first summarize what deep decarbonization planning entails and how

TABLE 6.1 Founding members of CNCA, with the North American members in bold.

Berlin, Germany	**Minneapolis, MN, USA**	Stockholm, Sweden
Boston, MA, USA	**New York City, NY, USA**	Sydney, Australia
Boulder, CO, USA	Oslo, Norway	**Vancouver, Canada**
Copenhagen, Denmark	**Portland, OR, USA**	**Washington, DC, USA**
London, UK	**San Francisco, CA, USA**	Yokohama, Japan
Melbourne, Australia	**Seattle, WA, USA**	

Source: based on USDN, 2015.

it is defined. CNCA positions members' deep decarbonization approach as transformative, in contrast to the incremental approach typically used to achieve 'interim' targets (e.g. GHG reduction of up to 20–30%). The significance of this is illustrated by a statement produced at the founding meeting of CNCA:

> It is possible to achieve many of the interim carbon reduction targets through continuous improvement in existing systems. But achieving [80% GHG reductions by 2050] will require transformative and systemic changes in many core city systems.
>
> *(CNCA, 2015, p. 125)*

Embracing a form of networked urban climate governance, this systemic approach is positioned by CNCA as a step forward from the piecemeal and incremental urban climate change response practices often characteristic of local governments. Yet, there are challenges to such urban climate governance. With local governments often lacking the required jurisdictional authority and financial autonomy, municipalities find that transnational climate change networks strengthen their authority and abilities through strategic partnerships (Gordon and Acuto, 2015). Such partnerships expand capacity and expertise, but can also limit the action taken by member cities since action is potentially influenced by accountability to other players (e.g. limiting action to approaches amenable to public–private partnership) (Gordon and Acuto, 2015). Furthermore, research on transnational urban climate networks has questioned the extensiveness of network impacts beyond a small number of pioneer cities (Kern and Bulkeley, 2009) and has pointed out that translation of network level activities to on-the-ground implementation is still an open question (Gordon and Acuto, 2015). The full impact and effectiveness of CNCA is yet to be realized. Yet, there are significant lessons to be drawn about the design of low carbon urbanism from deep decarbonization planning activities.

In 2015, CNCA provided insight into members' deep decarbonization planning in a document called *Framework for Long-Term Deep Carbon Reduction Planning* (CNCA, 2015). The work CNCA members did between approximately 2010 and 2015 to plan deep decarbonization fits into four key urban system categories: energy supply,

TABLE 6.2 Deep decarbonization planning strategy architecture of CNCA members.

Urban carbon emissions systems	Deep decarbonization strategies
Energy supply	• Decarbonize imported electricity • Increase local production of renewable power • Reduce demand for and consumption of electricity • Eliminate fossil fuel heating sources • Redesign models for utilities • Enable smart grids • Integrate city-wide energy management
Building energy efficiency	• Transform existing buildings into highly efficient and renewably powered structures • Incentivize and require net zero or renewable energy-positive new buildings • Increase the availability of building energy performance information in the marketplace • Advance/require performance-driven management of building energy • Grow the green buildings economic sector
Transportation	• Shift to a radically different mode share • Provide an array of modern, affordable, accessible mobility choices • Foster market dominance of clean technologies and fuels • Move quickly toward complete, connected, regionalized mobility systems • Change ways of thinking about and advancing alternative urban forms
Solid waste	• Get to zero waste • Promote sustainable consumption • Incentivize and require producer responsibility

Source: adapted from CNCA, 2015.

building energy efficiency, transportation and solid waste. Table 6.2 summarizes the strategies CNCA members plan to use to decarbonize urban systems.

Planning deep decarbonization also involves the institutionalization of carbon management via strategies such as creating accountability mechanisms, building technical capacity, engaging stakeholders, influencing other levels of government, funding climate action plans and stimulating innovation in government (CNCA, 2015). These strategies illustrate how deep decarbonization planning does not simply rely on the authority of the local government. Instead, the climate governance documents analysed mobilize a range of authorities and agencies in the pursuit of carbon neutrality, which demonstrates that overlapping scales and types of authority play a role in deep decarbonization planning.

Variations in ways of thinking about carbon can shape urban development. CNCA members concentrate their efforts under the banner of 'carbon neutrality'; yet, this notion is not as straightforward as it might seem at first sight. Consistency in the meaning of carbon neutrality cannot be assumed. CNCA members interpret the term in several different ways. For example, should carbon offsets be used to achieve carbon neutrality? Is natural gas expansion compatible with a carbon neutral future? Within CNCA, the answer to these questions has been both 'yes' and 'no' in different urban contexts. However, the underpinning logics guiding these decisions are likely to send cities down divergent socio-technical paths.

4 Low carbon as a tenet of urban development

The transition to net zero global carbon emissions requires transformations to energy systems, which implies substantial changes to the infrastructure systems that support life in cities (Rogelj et al., 2015). A comprehensive scope is important because parts of the climate change problem cannot be addressed in isolation. Climate change has been described as a 'wicked' problem, characterized by complexity and deep interconnections with material, social, political and economic aspects of society (Levin et al., 2012). Effective climate mitigation is only likely to be achieved by embedding climate responses into a broad range of policies at every level and, in this way, transitioning all societal systems to low carbon orientations (Burch et al., 2014; Loorbach, 2012; Levin et al., 2012). However, innovations in climate and energy policy to date do not appear to be adequate to achieve this dramatic transformation of communities (Burch et al., 2014). Instead, the transformation of urban systems requires the integration of new, low carbon logics as fundamental tenets of urban development.

Urban actors targeting decarbonization are trying to reshape urban development using these kinds of new low carbon logics. However, there are many different ways to be low carbon, which makes it important to closely examine decarbonization as it unfolds. As While et al. (2010) argue, there is no a priori reason why it should be assumed that carbon management would be used to meet progressive social goals. In fact, carbon management is a political project, and can be mobilized either to advance social justice or to further entrench inequalities (While et al., 2010; Bulkeley et al., 2013; Rice, 2014). This malleability in how carbon management is to be problematized, planned and implemented underscores why it is important to closely examine the new logics taking shape among pioneers of deep decarbonization.

The problem of carbon

There are many different ways to problematize carbon. CNCA members have declared a number of targets that can help us to understand the interpretation of carbon neutrality in each place. Figure 6.1 summarizes key targets for the North American members of CNCA.

Figure 6.1 shows that North American CNCA members have adopted several different logics to organize their GHG emission reduction efforts. Here, the bars

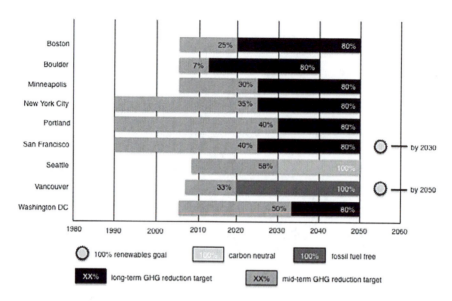

FIGURE 6.1 GHG and energy targets of North American founding members of CNCA. *Source*: adapted from CNCA, 2015.

illustrate the percentage of GHG emission reduction that municipal governments plan to achieve in their communities by a particular date. For example, Vancouver has goals to reduce GHG emissions 33% below 2007 levels by 2020, be fossil fuel free by 2050, and achieve 100% renewable energy by 2050. Like Vancouver, San Francisco also intends to achieve 100% renewable energy by 2050. Seattle plans to become carbon neutral. Boston, Minneapolis, New York City, Portland, San Francisco and Washington plan to achieve '80 × 50' carbon reductions (80% GHG reductions by 2050). However, the way that 80% is calculated is different: Boston plans 80% from 2005 emission levels (City of Boston, 2014); Minneapolis and Washington plan 80% from 2006 levels (City of Minneapolis, 2013; Washington D.C., 2012); New York City, Portland and San Francisco plan 80% from 1990 levels (City of New York, 2013; City of Portland, 2015; San Francisco, 2013); and Boulder plans to achieve 80% lower emissions from 2005 levels by 2040 (City of Boulder, 2015). In the end, GHGs will be reduced by different amounts in each place despite the shared overarching logic.

For a closer analysis, the following three brief vignettes highlight variation in logics underpinning the interpretation of carbon neutrality for buildings and energy in Vancouver, San Francisco and New York City. Note that transportation is outside of the scope of this chapter.

Vancouver

The pursuit of carbon neutrality in Vancouver is constituted by a fossil fuel free logic. Municipal-led climate governance policy documents do not specifically describe carbon neutrality, but are oriented around a goal to be fossil fuel free by 2040. Vancouver aims to be 100% renewable energy powered city-wide instead. In 2015, a total of 31% of Vancouver's city-wide energy use was from renewable sources (City of Vancouver, 2015b). Climate governance documents also outline a goal to reduce city-wide GHG emissions by 80% by 2050, but this goal is framed as inadequate on its own, established with the purpose of ensuring that the focus on renewable energy sources does not overlook GHG emission reductions (City of Vancouver, 2012). Vancouver aims to achieve fossil fuel free status using an integrated energy strategy that includes both demand- and supply-side measures. The Vancouver of the future is seen as relying on renewable energy sources, with a particular emphasis on replacing natural gas with renewable sources for both residential and district energy systems (City of Vancouver, 2015b). Buildings built across the city after 2020 are planned to be carbon neutral in operations (although the details of what this means are still vague) and the city is planning to increase urban density – further supporting possibilities for reducing emissions. Existing building energy use is planned to be reduced 20% from 2007 levels (City of Vancouver, 2012).

For Vancouver, fossil fuel-free status is to be achieved through a number of actions by a range of stakeholders. Residents, businesses, organizations and all levels of government are expected to be involved to make this vision a reality. For example, official documents stress that "Vancouver residents and businesses must be mobilized to become the agents for, and investors in, the changes that will come with a renewable city" (City of Vancouver, 2015b, p. 7). Building owners and developers will be requested to adopt efficient building construction techniques and sustainability retrofits to meet city mandates, and the City of Vancouver is planning to "develop innovative financing tools to help fund new zero-emission buildings" (City of Vancouver, 2015b, p. 30). Neighbourhood energy utilities are also a critical component of the planned fossil fuel-free strategy. Private companies are expected to operate these utilities and own low carbon energy infrastructure (City of Vancouver, 2015b). Pre-existing utilities will also continue to own electricity generation infrastructure, though the municipal government is planning to partner with them to increase the supply of renewable energy and to implement a smart grid (City of Vancouver, 2015b).

At the time of writing this chapter, implementation and progress was just beginning. In 2014, GHG emissions had been reduced 7% from 1990 levels (City of Vancouver, 2015a). Energy use across the city has been decreasing by 0.8% per year since 2007 (City of Vancouver, 2015b). In terms of energy infrastructure development, progress has been made in district energy implementation. In 2015, one neighbourhood energy utility served 4.2 million square feet of buildings with 60% energy from sewage waste heat recovery (City of Vancouver, 2015a). An Energy Retrofit Strategy was passed in 2014 and it is intended to support voluntary building efficiency improvements using energy benchmarking, energy audits and incentives. Increased energy efficiency requirements were included in the new building code and the re-zoning policy that took effect in 2014 and 2015 respectively (City of Vancouver, 2015a).

San Francisco

The logic guiding climate governance in San Francisco is based on a 100% renewables goal; 100% renewables for San Francisco means that, by 2030, residential electricity should come entirely from renewable sources and 80% of commercial electricity use (for industrial and business purposes) should come from renewable sources (San Francisco, 2013). In 2010, renewable energy provided 41% of the city-wide electricity mix for San Francisco (San Francisco, 2013).

Planned energy efficiency improvements complement this transition to renewables and, in particular, are expected to reduce natural gas use. Energy efficiency is to be improved through policies such as energy use benchmarking for large buildings and requirements for zero net energy (ZNE) residential building construction, where ZNE means "reducing energy use to the lowest level possible and then using renewable energy to supply the remaining energy needs on an annual basis" (San Francisco, 2013, p. 17). The former is planned to drive voluntary action by building owners, while the latter is an enforced standard. The City of San Francisco's goal with these efforts is "to increase transparency in building performance and incentivize the private market to increase energy efficiency through policies focused on market transformation" (San Francisco, 2013, p. 15). Furthermore, the City seeks to encourage energy demand reduction behaviour from building occupants across the city through education and awareness raising programs (San Francisco, 2013).

In addition to strategies that reduce emissions through improved efficiency and energy recovery, the City of San Francisco is working on "de-carbonizing the energy supply by replacing fossil fuels sources with renewable energy sources – micro-hydro, wind, geothermal, solar, wave, and biomass" (San Francisco, 2013, p. 12). Seventy-three per cent of the electricity used city-wide comes from an investor-owned utility and 16% from a utility owned by the City of San Francisco, with the remaining sourced from other energy service providers (San Francisco, 2013). The proposed mechanism of transition is purchases made by residents and businesses of San Francisco; customers are to be offered opportunities to purchase electricity from renewable sources from existing utilities, particularly the utility owned by the city. Renewable energy generation, particularly solar PV, is also to be supported by a number of local government policies (San Francisco, 2013).

There has been progress in implementation and GHG emissions reduction. In 2010, city-wide GHG emissions were 14.5% lower than 1990 levels, which was largely due to decreasing emissions intensity of electricity used in San Francisco because of the California's Renewables Portfolio Standard and the closure of two fossil fuel plants in San Francisco (San Francisco, 2013). San Francisco has adopted green building codes requiring LEED (Leadership in Energy and Environmental Design) Gold certification for new large commercial buildings and energy efficiency levels approximately 20% higher than the California Building Code (San Francisco, 2013). Mandatory building energy performance benchmarking and reporting has been in place since 2011 and has identified opportunities for 6 million

United States dollars (USD) in annual energy savings (San Francisco, 2013). The actual energy efficiency upgrades are not required. In addition, a City of San Francisco incentive program has provided 15.5 million USD to reduce the installation cost of 7 megawatts (MW) of solar PV, targeting residents, businesses and community organizations. This includes additional incentives for low-income residents and residents in 'environment justice' neighbourhoods that have historically experienced higher pollution levels (San Francisco, 2013, p. 22).

New York City

For New York City, deep decarbonization means achieving an 80% reduction in GHG emissions city-wide from 1990 levels by 2050, with a 30% reduction by 2030 as an interim target. According to climate policy documents produced by the municipality, achieving 80% by 2050 city-wide for New York City means substantial improvements in building energy efficiency (62% of planned reductions), cleaner power (18%), transportation (12%) and solid waste (8%) (City of New York, 2013). Energy efficiency of existing buildings is planned to be improved through substantial retrofits at an estimated cost of 4–5 billion USD a year for building owners, though it is argued that most of this investment would lead to operational cost savings over time (City of New York, 2013). Building owners and developers, it is argued, will need to switch sources of energy to technologies like geothermal, solar hot water, air source heat pumps, biogas, biomass CHP (combined heat and power), bio-diesel and conversion of heavy fuels to natural gas. In addition to retrofits, by 2050 new buildings are expected to be built 75% more efficiently than current construction standards (City of New York, 2013). The City of New York plans to spur these changes by providing technical support, improved access to financing, strengthening regulations, and through promotion (City of New York, 2013).

In 2013, two-thirds of New York's city-wide electricity supply was from natural gas, with the remainder sourced from nuclear and hydro. Though some inefficient generation will retire, generation located within the boundaries of New York City will be refurbished in the near future and will still operate in 2050 (City of New York, 2013). There are plans for an expansion in solar and wind power, as well as potentially the establishment of natural gas-fired cogeneration; similarly, hydro-power may be imported. The planned ownership of this infrastructure varies: "the City is working closely with utilities, research partners, and private businesses to accelerate the growth of clean distributed generation – including photovoltaic solar (PV) and combined-heat and power (CHP)" (City of New York, 2013, p. 8). Overall, research conducted by New York City finds that there is significant potential for zero carbon power:

> In theory, the technical potential that is available to New York City for zero-carbon resources is close to 30 GW [gigawatts], which would exceed existing installed capacity in the City even after de-rating capacity factors to account

for the intermittency of solar and wind resources. There are, however, significant and untested challenges to achieving this potential.

(City of New York, 2013, p. 64)

New York City's deep decarbonization pathway report cautions that the realization of this scope of zero carbon power is technically and economically difficult and, for these reasons, the report suggests that efficient natural gas generation will continue to have a role in the energy system in 2050 (City of New York, 2013).

In terms of implementation, New York City's city-wide GHG emissions fell by 19% between 2005 and 2012 (City of New York, 2013). Most of this reduction came from switching energy sources from coal and oil to natural gas. A package of building energy efficiency laws was passed in 2009, including mandatory energy and water use performance benchmarking for large buildings, retro-commissioning and some required upgrades of buildings long term (City of New York, 2013). Also, installed solar PV capacity across New York City has increased from 1 MW in 2007 to around 20 MW in 2013, although that represents just 0.2% of peak load (City of New York, 2013).

Pursuing decarbonization

The point here is not to learn from these vignettes as exemplars. Instead, it is about exploring the logics under development as urban low carbon transitions are designed and embedded. I draw three main points from these vignettes.

First, the vignettes show how carbon management goals are framed using terms such as 100% renewables, fossil fuel free, or carbon neutral. Though they are similar terms and share space under the CNCA's overarching carbon neutral frame, the variation in carbon control logics shows that different problematizations of carbon can lead to different socio-technical paths for cities. Vancouver problematizes fossil fuels entirely and seeks a complete transition to renewable energy. This logic leads to plans to reconfigure energy provision to increase neighbourhood-scale generation – as exemplified by the waste heat recovery project – and pursue increases in the integration of renewable energy in the energy mix provided by the main incumbent utilities. In San Francisco, fossil fuel use is not specifically problematized, so the 100% renewable energy goal is interpreted differently to focus on electricity as opposed to all energy and to source 80% of commercial electricity from renewable sources by 2030 rather than 100%. Investments in renewable energy will therefore focus on electricity provision for the urban energy system, with some room for continued fossil fuel use. The idea here is not to critique these targets, since the goals in both places represent a laudable integration of renewable energy. Instead, the point is that variations in the interpretation of carbon neutrality create different problems requiring different solutions, which ultimately affect the socio-technical nature of actual low carbon urbanism. The comparison with New York City is perhaps more stark.

A policy document for New York City produced by the municipality in consultation with other stakeholders declares a clean energy target that encompasses investment in new and refurbished natural gas generation (City of New York, 2013). Since the goal is an 80% reduction in emissions rather than zero carbon, this frame allows for the perpetuation of fossil fuel entrenchment up until 2050. Research conducted by New York City on a potential deep decarbonization pathway for the city argues that the generation capacity of the city can be more than replaced with technically available zero carbon energy sources (City of New York, 2013). However, the report found that apparently insurmountable concerns – costs, the intermittency of renewable energy sources, and the technical and financial viability of system integration of renewables – take a zero carbon goal off the table. Other cities using different logics see these same challenges as more manageable. If a fossil fuel-free logic is incorporated as a tenet of urban development instead, renewable energy would likely be encouraged and new investment in fossil fuel generation would likely be discouraged; the resulting urban energy systems would look different. In New York City, the specific logic framing how low carbon is understood leads to a different socio-technical path. Notwithstanding these important differences, there are still similarities in decarbonization paths despite the use of different logics: intense energy efficiency for buildings both new and old is seen as an important task in the pursuit of decarbonization regardless of the framing logic adopted.

Second, the vignettes illustrate the ways in which the problematization of carbon is impacted by existing socio-technical configurations woven into the materiality of the city. Different localities use different logics partly because cities have varied regional influences, historical development paths, and opportunities and challenges. Vancouver, for example, has the goal of 100% renewable energy. This goal fits well with a provincial electricity system that is already more than 90% renewable thanks to legacy hydropower development (BC Hydro, 2017). New York City, by comparison, has a power system two-thirds dependent on natural gas and with little renewable energy, which supports a logic that leaves space for natural gas entrenchment.

Third, the realization of low carbon urbanism impacts political configurations and power relations within the city. This is illustrated by the various ways in which low carbon infrastructure encounters different ownership modes. In San Francisco, for example, distributed ownership of solar PV systems among residents, businesses and community organizations is already under development. Equity in access to ownership opportunities is considered to be important since low-income residents are particularly supported by the municipality in renewable energy development. This is partly because lower income residents "spend a greater percentage of income on energy-related bills" (San Francisco, 2013, p. 20). Furthermore, solar PV ownership is being used as a tool to compensate victims of past environmental injustice. Maps were made of solar potential in targeted 'environmental justice' neighbourhoods, and residents in those areas have been offered additional incentives to pursue solar PV developments (San Francisco, 2013, p. 22). However, infrastructure ownership as a tool to address inequity is not a key feature of Vancouver and New York, and issues of ownership are not necessarily highlighted in key strategy documents. The fact that those who will own new energy infrastructure are likely to accrue economic benefits from the transition illustrates how power relations are affected by the integration of low carbon logics into urban development paths. This must be actively considered, both to avoid unintended consequences and to advance the progressive potential of low carbon urbanism. Without consideration of issues of power, equality and justice, moving towards low carbon urbanism may further entrench existing inequalities and miss out on any opportunities to more equitably redefine power relations in the city.

5 Objects, technology and infrastructure: elements of low carbon transformations

The previous section argued that different problematizations of carbon can lead to different socio-technical paths for cities, affecting the nature of actual low carbon urbanism. These socio-technical paths will be rendered through material objects, technologies and infrastructures, but numerous configurations of these elements could potentially constitute a low carbon city. Despite the fact that the low carbon city is often thought to be materially very different than the high carbon city, it is not clear exactly which elements will change. Therefore, this section examines the socio-technical elements that are being favoured in how a low carbon urbanism is imagined in North America.

The analysis of the climate governance frameworks of the seventeen founding members CNCA revealed sixteen objects, technologies and infrastructures as the key constitutive elements of the planned urban carbon neutral futures. These elements were identified using a discourse analysis approach, through a close reading of CNCA members' carbon governance policy documents (including documents such as carbon neutral strategy documents, climate change policy documents, and renewable energy strategies). Given common assumptions about carbon governance, a number of the elements are not a surprise – such as solar power, wind power, biofuels, hydropower, heat pumps, combined heat and power, and energy efficiency. Less tangible elements were included in the governance documents as well, such as reduced energy demand behaviour and spatial planning for low carbon (or 'climate districts'). The carbon governance documents also highlighted potentially surprising objects. Natural gas expansion, for example, might be an unexpected element of carbon neutrality since it represents further entrenchment of fossil fuels. All sixteen elements are shown in Figure 6.2. While we might think about some of these as technology interventions (e.g. solar power) and about others as behavioural changes (e.g. reduced energy demand), it is more productive to consider all these elements from a socio-technical perspective: they combine materials and technologies as well as ways of doing and other social aspects. They are a mixture of objects and agencies, wrapped in a political context.

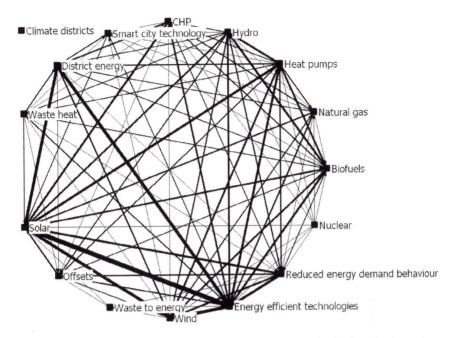

FIGURE 6.2 Visualization of the key socio-technical elements highlighted in the carbon governance policy documents of the nine North American founding members of CNCA.

Note: The nodes represent the urban elements related to the building and energy sector that will be mobilized to achieve carbon neutrality. The connections (lines) show that the elements co-occur in carbon governance texts and the relative frequency of co-occurrence is shown using line weight.

Source: the author.

According to the municipal-led carbon governance policy documents examined, low carbon cities in North America are likely to be composed of these elements and the relationships among them. One way to consider the relationships between these elements of low carbon transformations is to use textual network analysis methods. Textual network analysis identifies connections that exist between ideas or arguments across a sample of texts (Klenk and Larson, 2013; Farrell, 2015). As an analytical tool, it can be used to reveal the rhetorical structure of discourses and to help clarify meaning within texts. Figure 6.2, drawn based on an analysis of the imaginaries of low carbon urbanism embedded within the nine cities studied, shows the connections that exist between the various urban elements that are planned to be mobilized to achieve urban carbon neutral futures. This visual analysis represents links between urban elements in order to understand which socio-technical combinations dominate urban carbon governance plans. The analysis, visualized using Netdraw, reveals the number of times that particular urban elements were mentioned within the carbon governance documents analysed. It illustrates the relative importance of various combinations of urban elements to how the low carbon cities of the future are currently imagined in North America. The nodes of the visualization represent elements that are planned to be mobilized in order to achieve carbon neutrality. Lines between elements mean that they co-occur in the governance texts for a particular city. A thicker line means that those objects co-occur more frequently across the studied sample of texts (e.g. solar and energy-efficient technologies frequently co-occur). The most frequently occurring elements appear at the heaviest confluence of line weights (e.g. energy-efficient technologies). Accordingly, the elements at the heaviest confluence of line weights represent the dominant configurations that are planned to be mobilized to achieve urban carbon neutrality in North America.

Some elements are clearly more dominant in the visualization than others, which means that, in the nine studied cities, they are mobilized more often in imaginaries of low carbon urbanism. Solar power and energy-efficient technology appear to be considered particularly key components of urban carbon neutral futures. District energy, heat pumps, hydropower, wind power, biofuels and reduced energy demand behaviour are also frequently mobilized. Mobilized less often are smart city technologies, CHP (combined heat and power) and natural gas. The least emphasized (though still included) elements are nuclear power, waste heat, offsets and waste-to-energy. The climate districts element has no connections because it was not an emphasized element for any of the North American cities. It is included in the figure because it was an element that is going to be mobilized to achieve carbon neutrality among the larger set of seventeen members of CNCA.

What does this tell us about how low carbon cities are imagined? Figure 6.2 reveals an emerging picture of a city composed of highly energy-efficient buildings powered by solar energy while also using a district heating approach. A suite of renewable fuels and technologies are also employed. Behaviour change to reduce energy demand is emphasized. Natural gas expansion, perhaps employed in combined heat and power applications, is not discounted, and waste heat, waste-to-energy and nuclear technologies have minor roles to play. While this pattern of low carbon urbanism emphasizes highly energy-efficient buildings powered by solar energy using a district approach, some European members of CNCA are aiming for an urban future much more heavily dependent on district energy using biomass-powered CHP and a compact built environment (Tozer and Klenk, submitted). This illustrates how the materiality and socio-political context of different urban geographies impacts how low carbon is being embedded into urban development. Therefore, drawing together the argument from the previous section, the socio-technical nature of low carbon urbanism is influenced both by the logics shaping ways of thinking about carbon and the materiality of urban geographies.

6 Conclusions

This chapter has examined how the idea of 'carbon neutrality' within the building and energy sector is represented in nine local North American authorities striving for deep decarbonization. As this analysis has shown, diverse interpretations of carbon neutrality create different problems requiring different solutions, which influences the nature of actual low carbon urbanism. These variations impact the socio-technical development path of the city. Carbon governance based on different logics – such as 100% renewables, fossil fuel free and 80% by 2050 reduction targets – will have different implications for the transformation of urban systems. On one hand, variation in low carbon logics is largely positive since it allows decarbonization to be customized to suit local context. On the other hand, this variation could serve to hide the further entrenchment of fossil fuels under a 'carbon neutrality' label, and move urban design away from decarbonization.

There are many different ways for urban systems to be low carbon, but particular socio-technical interventions are being favoured in how low carbon urbanism is imagined in North America. When considering what kind of urban future is in store, the materiality and socio-political context of a place matters. Furthermore, although this chapter has shown that imaginaries of carbon neutrality consistently mobilize a broad swathe of socio-technical elements, different kinds of carbon neutral urban futures are envisioned for different places. Ways of thinking about carbon are impacted by the existing regional influences, historical development paths, and opportunities and challenges. This was clear in the vignettes describing plans to decarbonize buildings and energy systems in Vancouver, San Francisco and New York City. The patterns of mobilization for low carbon elements in the cities evaluated suggest that low carbon is being embedded into urban development differently across urban geographies, illustrating how the problematization of carbon is impacted by existing socio-technical configurations embedded in the materiality of the city.

Among the North American founding members of the Carbon Neutral Cities Alliance, low carbon urban planning and design is influenced both by the logics shaping ways of thinking about carbon and the materiality of urban geographies. However, deep decarbonization is still in the planning stages, and therefore its implications are not yet clear for urban development in practice. The ways that the design of carbon neutral urbanism is practised and mobilized will affect the achievement of transformation in the future.

References

BC Hydro (2017) *Generation System*. Vancouver: BC Hydro. Available at: https://www.bchydro.com/energy-in-bc/our_system/generation.html (accessed 21 April 2017).

Bernstein, S. and Hoffmann, M. (2016) *The Politics of Decarbonization: A Framework and Method*. Environmental Governance Lab Working Paper. Toronto: University of Toronto.

Betsill, M. and Bulkeley, H. (2007) Looking back and thinking ahead: a decade of cities and climate change research. *Local Environment*, 12(5), 447–456.

Bulkeley, H., Andonova, L.B., Betsill, M.M., Compagnon, D., Hale, T., Hoffmann, M.J., Newell, P., Paterson, M., Roger, C. and VanDeveer, S.D. (2014) *Transnational Climate Change Governance*. New York: Cambridge University Press.

Bulkeley, H. and Betsill, M.M. (2013) Revisiting the urban politics of climate change. *Environmental Politics*, 22(1), 136–154.

Bulkeley, H., Carmin, J., Castán Broto, V., Edwards, G.A.S. and Fuller, S. (2013) Climate justice and global cities: Mapping the emerging discourses. *Global Environmental Change*, 23(5), 914–925.

Bulkeley, H., Castán Broto, V. and Maassen, A. (2011) Governing urban low carbon transitions. In: Bulkeley, H., Castán Broto, V., Hodson, M. and Marin, S. (eds.) *Cities and Low Carbon Transitions*. New York: Routledge, pp. 29–41.

Burch, S., Shaw, A., Dale, A. and Robinson, J. (2014) Triggering transformative change: a development path approach to climate change response in communities. *Climate Policy*, 14(4), 467–487.

Castán Broto, V. and Bulkeley, H. (2013) Maintaining climate change experiments: urban political ecology and the everyday reconfiguration of urban infrastructure. *International Journal of Urban and Regional Research*, 37(6), 1934–1948.

City of Boston (2014) *Greenovate Boston 2014 Climate Action Plan Update*. Available at: https://www.cityofboston.gov/eeos/pdfs/Greenovate%20Boston%202014%20CAP%20Update_Full.pdf (accessed 17 July 2017).

City of Boulder (2015) *Boulder's Climate Commitment: Rising to the Climate Challenge, Powering a Vibrant Future (Draft)*. Boulder: City of Boulder. Available at: https://www-static.bouldercolorado.gov/docs/Boulder_Climate_Commitment_Doc-1-201510231704.pdf (accessed 17 July 2017).

City of Minneapolis. (2013). *Minneapolis climate action plan: A roadmap to reducing citywide greenhouse gas emissions*. Minneapolis: City of Minneapolis. Available at http://www.minneapolismn.gov/www/groups/public/@citycoordinator/documents/webcontent/wcms1p-109331.pdf (accessed 17 July 2017).

City of New York (2013) *PlaNYC: New York City's Pathways to Deep Carbon Reductions*. New York: City of New York. Available at: http://s-media.nyc.gov/agencies/planyc2030/pdf/nyc_pathways.pdf (accessed 17 July 2017).

City of Portland (2015) *Climate Action Plan: Local Strategies to Address Climate Change*. Portland: City of Portland. Available at: https://www.bbhub.io/mayors/sites/14/2015/06/Portland-Action-Plan.pdf (accessed 17 July 2017).

City of Vancouver (2012) *Greenest City 2020 Action Plan*. Vancouver: City of Vancouver. Available at: http://vancouver.ca/files/cov/Greenest-city-action-plan.pdf (accessed 17 July 2017).

City of Vancouver (2015a) *Greenest City 2020 Action Plan Part Two: 2015–2020*. Vancouver: City of Vancouver. Available at: http://vancouver.ca/files/cov/greenest-city-2020-action-plan-2015-2020.pdf (accessed 17 July 2017).

City of Vancouver (2015b) *Renewable City Strategy*. Vancouver: City of Vancouver. Available at: http://vancouver.ca/files/cov/renewable-city-strategy-booklet-2015.pdf (accessed 17 July 2017).

CNCA (Carbon Neutral Cities Alliance) (2015) *Framework for Long-Term Deep Carbon Reduction Planning*. Denmark: CNCA. Available at: https://www.usdn.org/uploads/cms/documents/cnca-executive-sum_web.pdf (accessed 17 July 2017).

CNCA (Carbon Neutral Cities Alliance) (n.d.) *Background*. CNCA website. Available at: https://www.usdn.org/public/page/75/Background (accessed 7 June 2017).

Farrell, J. (2015) *The Battle Over Yellowstone: Morality and the Sacred Roots of Environmental Conflict*. Princeton, NJ: Princeton University Press.

Gordon, D.J. and Acuto, M. (2015) If cities are the solution, what are the problems? – the promise and perils of urban climate leadership. In: Toly, N.J., Johnson, C. and Schroeder, H. (eds.) *The Urban Climate Challenge: Rethinking the Role of Cities in the Global Climate Regime*. New York: Routledge, pp. 63–81.

IPCC (Intergovernmental Panel on Climate Change) (2015) Human settlements, infra-structure, and spatial planning. In *Climate Change 2014: Mitigation of Climate Change; Working Group III Contribution to the IPCC Fifth Assessment Report*. Cambridge: Cambridge University Press, pp. 923–1000.

Kern, K. and Bulkeley, H. (2009) Cities, Europeanization and multi-level governance: governing climate change through transnational municipal networks. *Journal of Common Market Studies*, 47(2), 309–332.

Klenk, N.L. and Larson, B. (2013) A rhetorical analysis of the scientific debate over assisted colonization. *Environmental Science and Policy*, 33, 9–18.

Levin, K., Cashore, B., Bernstein, S. and Auld, G. (2012) Overcoming the tragedy of super wicked problems: constraining our future selves to ameliorate global climate change. *Policy Sciences*, 45(2), 123–152.

Loorbach, D. (2012) *Transition Management: New Mode of Governance for Sustainable Development*. Utrecht: International Books.

Lovell, H., Bulkeley, H. and Owens, S. (2009) Converging agendas? Energy and climate change policies in the UK. *Environment and Planning C: Government and Policy*, 27(1), 90–109.

Reckien, D., Flacke, J., Dawson, R.J., Heidrich, O., Olazabal, M., Foley, A., Hamann, J.J.P., Orru, H., Salvia, M., De Gregorio Hurtado, S., Geneletti, D., and Pietrapertosa, F. (2013) Climate change response in Europe: what's the reality? Analysis of adaptation and mitigation plans from 200 urban areas in 11 countries. *Climatic Change*, 122(1–2), 331–340.

Rice, J.L. (2014) An urban political ecology of climate change governance. *Geography Compass*, 8(6), 381–394.

Rogelj, J., Luderer, G., Pietzcker, R.C., Kriegler, E., Schaeffer, M., Krey, V. and Riahi, K. (2015) Energy system transformations for limiting end-of-century warming to below 1.5 °C. *Nature Climate Change*, 5, 519–527.

Romero-Lankao, P. (2012) Governing carbon and climate in the cities: an overview of policy and planning challenges and options. *European Planning Studies*, 20(1), 7–26.

San Francisco (2013) *Climate Action Strategy 2013 Update*. San Francisco: City of San Francisco. Available at: https://sfenvironment.org/sites/default/files/engagement_files/sfe_cc_ClimateActionStrategyUpdate2013.pdf (accessed 17 July 2017).

Scoones, I., Leach, M. and Newell, P. (2015) The politics of green transformations. In: Scoones, I., Leach, M. and Newell, P. (eds.) *The Politics of Green Transformations*. New York: Routledge, pp. 1–23.

Toly, N.J. (2008) Transnational municipal networks in climate politics: from global governance to global politics. *Globalizations*, 5(3), 341–356.

Tozer, L. and Klenk, N.L. (Submitted) Urban configurations of carbon neutrality: insights from the Carbon Neutral Cities Alliance.

Unruh, G.C. (2000) Understanding carbon lock-in. *Energy Policy*, 28(12), 817–830.

USDN (Urban Sustainability Directors Network) (2015) Carbon Neutral Cities Alliance. USDN website. Available at: https://www.usdn.org/public/page/13/CNCA (accessed 21 April 2017).

Washington D.C. (2012). *Sustainability DC: Sustainable DC Plan*. Washington D.C.: District of Columbia. Retrieved from https://sustainable.dc.gov/sites/default/files/dc/sites/sustainable/page_content/attachments/DCS-008 Report 508.3j.pdf (accessed 17 July 2017).

While, A., Jonas, A.E.G. and Gibbs, D. (2010) From sustainable development to carbon control: eco-state restructuring and the politics of urban and regional development. *Transactions of the Institute of British Geographers*, 35(1), 76–93.

PART II

Intermediation and governance

7

RECONFIGURING SPATIAL BOUNDARIES AND INSTITUTIONAL PRACTICES

Mobilizing and sustaining urban low carbon transitions in Victoria, Australia

Susie Moloney and Ralph Horne

1 Introduction

Carbon and efforts to decarbonize are reconfiguring urban processes and relations. As we witness increasing numbers and ranges of low carbon urban experiments attention inevitably turns to how they are sustained (Castán Broto and Bulkeley, 2013a) and how, as part of this process of sustaining, urban low carbon intermediaries (Guy et al., 2011; Hodson et al., 2013) at the local scale operate across existing boundaries, between civil society, policy and the private sector, in deliberative ways.

In this chapter we develop earlier work (Moloney et al., 2010; Horne and Dalton, 2014; Moloney and Horne, 2015; Moloney and Funfgeld, 2015) focusing on the role of emerging quasi-government networks operating as intermediaries within and between not-for-profit, government and business organizations and across spatial and jurisdictional boundaries. We propose that the ways in which these types of organizations create spaces for experimentation across local government boundaries – through projects, relations and strategies – can be understood as intermediation, and that this is particularly important work in weak institutional settings, where climate change policy is contested and interests are distributed. It is in this highly contested and fragmented domain of low carbon governance that we identify the characteristics of their work as constituting forms of intermediation (Hodson and Marvin, 2010) by

- coordinating the development of regional climate change plans and strategies;
- establishing progressive coalitions seeking to transform socio-technical systems;
- advocating for coordinated, systemic and effective responses to climate change across multiple scales; and
- seeking to align and legitimize low carbon agendas across local governments, businesses and civil society in their particular regions.

We start from the premise that there are two directions in which local governance intermediation is required; vertically, i.e. between levels of government and at different scales; and horizontally, i.e. across local spatial boundaries, organizations, networks and stakeholders and their interests. Much has been said about the vertical dimensions of this phenomenon in the literature on multilevel governance (Bulkeley and Betsill, 2005). Here, we turn our attention in particular to the horizontal dimensions and, within this, the less often studied ways in which particular intermediaries seek to link, frame, create and sustain new low carbon projects and activities across local and regional scales.

A 2015 global Urban Climate Change Governance survey of 350 local authorities highlighted the importance of local government networks both *within* and *across* governments and identified the need for a clearer understanding of internal network building and the role of formal and informal approaches to networking and institutionalization (Aylett, 2015). More recently, a study of local climate policy and action in small to medium cities highlighted the importance of 'inter-municipal networking', particularly where there was already a history of network collaboration between municipalities as opposed to larger scale cities whose collaborations tend to be with peers, national and international partners (Hoppe, 2016).

This case study focuses on the Climate Change Alliances (CCAs) in Victoria, Australia. They are formed by voluntary agreements between groups of adjoining local governments (also called local authorities, or councils; the base level of the Government in Australia). Hence, each CCA covers the area of several local authorities. They are quasi-government networks in as much as they are not formally government entities, but they are controlled primarily by a steering committee of local government representatives. CCAs are charged with a broad agenda of developing low carbon projects, plans and programmes and sharing practices, knowledge and networks. They are typically funded on a shoestring by year-to-year local agreements to co-fund a single position or a small office. They vary in terms of legal standing, from informal non-legal entities to more formally constituted enterprises.

There are 10 alliances in the state of Victoria, Australia, involving 72 of 79 municipalities, each comprising between 4 to 8 local councils and other not-for-profit organizations. We focus on their roles in leading and coordinating regional-scale mitigation and adaptation planning and responses, capacity-building within and across local government boundaries and facilitating partnerships across civil society, business and government sectors and other local actors. We examine the work of a particular alliance in the western region of Melbourne as they seek to challenge existing policy and institutional arrangements governing 'business as usual' development modes. They have enlisted local governance organizations, utilities and industry in their visioning process, mobilizing and connecting a range of actions across the region. We examine these processes of transitioning and analyse to what extent this contributes to a sustained low carbon development transition.

These activities in Victoria are occurring against a contested, partisan political national landscape. Vertically, governments are not aligned on carbon, as highlighted by the controversial decisions at the federal level by the Abbott Conservative government in 2014 to repeal the national Carbon Pollution Reduction Scheme (CPRS) – introduced by the previous Labour government – to weaken the renewable energy target (previously 20% by 2020), and to replace the polluter pays model of the CPRS with an Emissions Reduction Fund. In the lead-up to the 2015 United Nations Climate Change Conference, COP21, in Paris, Australia's emissions reduction target (26–28% below 2005 levels by 2030) and climate policy was considered 'one of the weakest and more ineffective in the developed world' (Schlosberg, 2015). Meanwhile, at the state level in Victoria, a similar political shift and rejection of climate change science occurred under a Conservative government, replacing the words 'climate change' with 'climate variability' in policy documents and repealing a legislated emissions target. This was followed in November 2014 by the return of a Labour government with a stated commitment to positioning Victoria as a leader in climate mitigation and adaptation. In February 2017 the Victorian State Government created a new Climate Change Act, which includes the establishment of an emissions reduction target of net zero by 2050, five yearly interim targets, a set of policies and principles to embed climate change in government decision-making and a requirement for the government to develop a climate change strategy every five years (Parliament of Victoria, 2017). This sits alongside the release in 2017 of a Victorian Climate Change Framework, Adaptation Plan (2017–20) and Renewable Energy Action Plan. While this represents an important shift in state government climate change policy in recent years the implementation of many initiatives are at a very early stage.

In this volatile and uncertain climate change policy context, and despite the relatively financially and politically weak role of local governments in Australia, they have emerged as key actors involved in a broad range of low carbon interventions incorporating behaviour/social change; retrofitting/demonstration; transition and advocacy (McGuirk et al., 2014). As a product of and vehicle for these efforts, CCAs have filled an institutional and policy gap that has opened up in climate change governance between the politically and financially weak fragmented local government scale and the inconsistent and short-termist state and federal scales.

A recent national survey of Australian local government carbon reduction initiatives positions them as "experiments that are exploring institutional partnerships to expand governance capacity and authority" (McGuirk et al., 2015, p. 50) with the potential to shape urban political and socio-technical transformations. While the role of 'cross border' institutional and governing arrangements has been highlighted as important in addressing complex natural resource management issues, there remains limited research around how these arrangements are emerging to address climate change (Steele et al., 2013, p. 700) particularly across local authority boundaries. In this chapter we examine inter-local-government networks and their role in reconfiguring climate change responses from local to regional scale through mobilizing formal and informal institutional practices and relations. We focus our attention on three key questions:

1.   Where and how have CCAs emerged in Melbourne (and Victoria)?
2.   How do CCAs build governance capacity?
3.   How can CCAs work in mobilizing change be rendered legitimate and sustained as transition activities?

Here we take 'legitimacy' to mean the process of becoming accepted as a bona fide actor, exercising power and influence, contributing to capacity-building, and acting as an acknowledged and appropriate vehicle for innovation and relationship-building. In this regard, the extent to which the CCAs are rendered legitimate is important in the shaping of low carbon urbanism as a sustained normative goal-driven strategic urban and economic development process.

2 Modes of local governance and the role of intermediaries

Previous work at the subnational scale and in particular of local governments in local climate action has explored the multiple modes of governing (Bulkeley and Betsill, 2003) and forms of experimentation (Hoffman, 2011; Castán Broto and Bulkeley, 2013b) emerging largely in response to inaction or a lack of progress at international and national scales. Such experiments include both project innovation and governance innovation. Modes of governing at the local scale have been characterized in four categories: governing by authority, governing by provision, self-governing and governing by enabling (Bulkeley and Kern, 2006). Hoppe (2016) discerned a fifth governance mode characterized as horizontal, networked local climate change governance, which it is argued complements the earlier four categories.

In general, local climate change governance experimentation seeks to (re)make rules that shape how communities respond to climate change, that are independent of international-scale climate negotiations and agreements, and that traverse (local) jurisdictional boundaries. Since experimentation invokes dynamics and boundaries being crossed, it follows that intermediaries that transact across these boundaries and through these experiments are potentially important in emerging modes of governance and forms of restructuring at the local scale. They must engage with social, economic and technical processes in what has been described as niche experiments operating as "the locus of radical innovations" (Geels, 2005, p. 451). Geels contends that niches are important because they provide "space for learning processes on several dimensions" (e.g. technical, user preferences, regulations, infra-structures and symbolic meaning) and "space to build social networks that support innovations" (e.g. lobby groups, user associations and new industry networks) (2005, p. 451). While this is useful for our analysis, it lacks a spatial dimension that we regard as necessary to understand local climate change governance in action.

To address this gap in transitions studies, we join other emerging work on urban/spatial transitioning in Australia (Moore et al., 2017) in proposing situated spatial studies of low carbon transitions in order to reveal the socio-materialities at play. Processes of transitioning involve intermediation across physical terrains as well as across production and consumption; actors, scales, priorities, interests and funders; and plans, priorities and their application (Hodson et al., 2013). Intermediary characteristics and activities include the development of plans and strategies augmented by funding, partnerships and coalitions (Hodson and Marvin, 2010) which can be further grouped into four mediator categories (Fischer and Guy, 2009, p. 2588):

1. bridge-builders, mediators, go-betweens or brokers, facilitating dialogues, resolving conflicts or building partnerships;
2. 'info-mediaries', disseminating information, offering training and providing technical support;
3. advocates, lobbyists, campaigners, gatekeepers or image-makers, fighting for particular causes; and
4. commercial pioneers, innovators and 'eco-preneurs'.

Considering low carbon urbanism as a contested political and spatial project, in our analysis we draw also on the work of urban planning theorists and see an alignment with the notion of 'progressive place governance' as described by Healey (2015). Place-based experiments can constitute an "urban politics that can be engaged for progressive ends," and the governing of carbon can be regarded as a means of "achieving other forms of social, economic and environmental benefits for those who have previously been excluded" (Bulkeley et al., 2014, p. 254). The role of experimental innovation is recognized in progressive place governance alongside the need for strategic judgement, legitimacy and public value (Healey, 2015, p. 24). A key challenge for the future is "how we can make an effective relationship between state support and civil society" (Healey, 2015, p. 24). In the case of low carbon urban governance, key questions emerge around the purposive and progressive capacity of low carbon experiments to mobilize change in institutional arrangements, practices and norms involved in "making, maintaining and living" low carbon urban futures (Bulkeley et al., 2014, p. 49).

In framing our analysis of CCAs as intermediaries we recognize their actions as constituting a form of 'progressive place governance' (Healey, 2015). In taking a place-based lens to processes of transitioning we highlight in our analysis what we regard as the essential connections between the 'where' of low carbon urbanism as well as the 'how'. The following draws on document analysis and interviews with the chief executives (and one project officer) of nine of the ten CCAs. The interviews were semi-structured and in-depth and were carried out between October 2014 and August 2015. These interviews were designed to explore the modus operandi of the CCAs, specifically: their business model and networks; and their capabilities, skills and learning processes.

3 The development of CCAs

In this section we address the question: where and how have CCAs emerged in Melbourne (and Victoria)? The alliances originally formed as informal 'information-sharing' networks among local government sustainability officers largely across adjacent local authorities. These networks were facilitated more formally through a 2002 Regional Partnerships Program as part of the then Labour state government's Victorian Greenhouse Strategy. The need to improve capacity across local governments was recognized through building networks and partnerships. The program provided the salary for each alliance to fund an executive officer whose task was to coordinate and mobilize partner council's mitigation initiatives. In the mid- to late 2000s the priorities were street lighting and energy efficiency programs focusing on council-owned buildings. The Regional Partnerships Program was reviewed in 2006 and while it was deemed a success in achieving the goals of capacity-building, collaboration and project implementation, the state government ceased providing its annual contributions to alliances in 2008. While there was a period of transition for some of the alliances, most partner councils agreed to provide a small annual fee to sustain the role of the executive officer and ensure the continuity of initiatives and progress in regional strategies.

Of the ten alliances in Victoria, those covering the Melbourne Metropolitan area have been more adequately resourced, and perhaps because of proximity, they have maintained stronger networks, developed in most cases over a period of many years. The metropolitan-based alliances have had more success in mobilizing climate change planning and projects than their regional- (largely rural-)area counterparts. The four metropolitan alliances covering all but six of Melbourne's local governments are particularly active; the Western Alliance for Greenhouse Action (WAGA), Northern Alliance for Greenhouse Action (NAGA), Eastern Greenhouse Alliance and South Eastern Councils Climate Change Alliance (SECCCA) (see Table 7.1). Each alliance radiates spatially along four corridors and includes inner higher density zones through to low density urban fringe suburbs as well as rural and coastal peri-urban areas. The alliance areas collectively include 3.8 million residents; almost the entire population of Greater Melbourne. They are comparable in terms of the diversity of communities they work with (i.e. demographically, socio-economically, culturally and politically) and land-use challenges they face (i.e. including variations in infrastructure, in built form, quality of housing stock, interface between urban and rural areas, etc.). The spatial pattern of CCAs may be explained by a historical trend in Melbourne involving urban planning along similar corridors initially following the pattern of rail and road transport infrastructure routes. Urban planning strategies since the post-war era have reinforced the continued expansion of Melbourne along these corridors, and in adjacent 'growth areas' earmarked due to suitable land availability.

While regional-scale networks are not unusual in Australia, for example in natural resource management and regional economic development, the emergence of regional-scale CCAs is unique to Victoria. As an 'informal' tier of regional governance, CCAs have played an important role in mediating both horizontally across local government boundaries and institutional practices and vertically between groups of councils (regionally) and state and federal governments, and increasingly with civil society, businesses and non-governmental organizations. The role of the executive

TABLE 7.1 Melbourne Metropolitan CCAs.

Metropolitan alliances	Year established	Current members	Population	GHG emissions by sector (2013 est.)
WAGA	2006	8 local governments	1.02 million	Buildings. 25%
				Transport, 12.5%
				Freight, 12%
				Industry, 48%
NAGA	2002 (formally 2006)	9 local governments; not-for-profit Moreland Energy Foundation	958,000 million	Buildings, 43%
				Transport, 21%
				Freight, 6%
				Industry, 30%
				Agriculture, water, waste, 5%
EAGA	2008 (2012 formally)	6 local governments	1 million	—*
SECCCA	2004	8 local governments	820,000 million	Industry, 39%
				Residential, 23%
				Transport, 22%
				Commercial, 11%
				Agriculture, 3%
				Waste, 2%

* Emissions not reported.

Note: EAGA, Eastern Alliance Greenhouse Action; GHG, greenhouse gas; NAGA, Northern Alliance for Greenhouse Action; SECCA, South East Councils Climate Change Alliance; WAGA, Western Alliance for Greenhouse Action.

Source: adapted from AECOM and Arup, 2013.

officer is critical in maintaining their work and developing the leadership potential of alliances to shape agendas across their member local authorities and partners as well as at state government levels. Meeting regularly, CCA executive officers share strategies, learning and coordinate advocacy processes to lobby state policy-makers and regulators, which is an important process in both sustaining and building their capacity, which we will focus on in the next section.

4 Building governance capacity

In this section we address the question: how do CCAs build governance capacity? Within each CCA member local governments contribute an annual membership fee in the order of 15,000 Australian dollars (AUD), which funds the salary of an executive officer and some operational expenses. While each has a different governance structure, all include an executive officer and most have an executive committee (constituted by local government directors and sometimes elected councillors) and an operational committee (often made up of sustainability officers). SECCCA uniquely chose to become an 'incorporated association', affording it a degree of independence from its member local governments, giving it the capacity to apply for grants and act as a quasi-independent consultancy:

> we can enter into contracts on our own basis … if you've got eight councils, you can only move as fast as your slowest one … And with eight councils, there's one likely to be quite slow … So we can operate in our own right.
>
> *(SECCCA Executive Officer, October 2014)*

Each alliance has a Memorandum of Understanding with their member local governments, which is reviewed every three to five years. This means that annual contributions and membership can be 'fluid', with both up for discussion during periods of renewal. So, alliances are accountable to member local governments for the value of ongoing voluntary membership fees:

> every time we did our annual strategic planning session, we have to answer – why does SECCCA need to exist for another year? … We don't have a God given right to exist, we only exist so long as we're doing something of worth to them, and if we can't answer that question in the affirmative, see you later.
>
> *(SECCCA Executive Officer, October 2014)*

While originally the alliances were set up to 'do projects' (SECCCA Executive Officer, October 2014), they have over time adapted and taken on more strategic roles, developing regional risk assessments, greenhouse strategies and adaptation plans largely facilitated through gaining federal and state government grants. This capacity to attract external funding (in SECCCA's case a total of 9 million AUD between 2004 and 2015 as compared to 1 million AUD in membership fees over the same time period) has been important in proving their value to member local governments:

> The reason why we've been able to get them to commit to active participation in an alliance is we say: 'We will bring in some external money'.
>
> *(SECCCA Executive Officer, October 2014)*

A key function of CCAs is the ability to leverage funds from other higher tiers of government to facilitate cross-municipal networking and capacity-building. While this capacity-building role continues to be central to their work, they are increasingly seeking funding opportunities to strengthen their strategic role through the development of long-term regional plans:

> Well as I said, it started off as really a collection of sustainability officers getting together to see what collaborative projects were possible, but also just to share information, and to do some capacity-building amongst themselves. And

WAGA still definitely has that role, and that is still a very important role. But we have, as I said, we've just developed a regional greenhouse strategy which addresses mitigation. We also have a regional adaptation strategy. So we're actually becoming a lot more strategic.

(WAGA Executive Officer, October 2014)

5 Building legitimacy and sustainability

In this section we address the question: how can CCAs work in mobilizing change be rendered legitimate and sustained as transition activity? Healey (2015, p. 20) describes institutional capacity-building as involving processes "developed through the interplay of knowledge resources and relational resources." These processes she argues generate "mobilization capacity," which is the capacity to "keep doing new things in new ways." This can also be understood as experimenting with new ways of doing that necessitates reframing and reconfiguring meanings, materials and competencies (Shove et al., 2012) within local government institutional practices and in how they relate with other actors in their particular contexts. While experimenting with new projects and strategies, place-based organizations and other 'informal' governance arrangements must maintain legitimacy and account-ability to their stakeholders and communities. In the case of CCAs this includes being accountable to their member councils, to ensure they continue to provide 'added value', but also in how they help mobilize 'public value' – which Moulaert (2010) argues is a critical dimension of social innovation. While CCAs must render themselves legitimate through achieving outcomes and outputs, they also develop legitimacy through their 'discursive power' in shaping agendas, advocacy for councils and communities, and creating 'new ways of thinking' for example around low carbon economic development agendas. The regional 'voice' offered by CCAs is considered to carry more political weight strengthening the legitimacy of local concerns particularly to state and federal governments. As stated by the Project Officer from NAGA, "NAGA could say things that sometimes they (councils) couldn't say and that it had a respected voice" (NAGA Project Officer, October 2014).

Alliances through their advocacy work not only seek to influence structural processes, but also enlist a range of actors and agents into their strategic work:

Regional-scale urgency, you know that sense, let's scale things up, let's work out what works and beef it up and get it to happen everywhere, as much as we can. Advocacy I've mentioned and capacity-building which is again the role of the alliance, the network or bringing the new players and the sharing of knowledge and capability.

(NAGA Project Officer, October 2014)

An example of an alliance attempting to experiment and reconfigure economic development agendas is WAGA's Low Carbon West (LCW) strategy development.

WAGA represents the western wedge outwards from greater Melbourne into regional Victoria. This is the fastest growing region of Australia with population expected to grow from 1.02 million to 1.46 million by 2031 and an additional 180,000 new homes largely in greenfield locations. LCW was developed by WAGA with project partners LeadWest and Regional Development Australia (RDA), Western Melbourne. LeadWest (which is constituted as a company limited by guarantee) includes six local governments (also members of WAGA) and involves major companies and other organizations each with substantial operations and interests based in Melbourne's west. AECOM and Arup were jointly commissioned as the project consultants and led the consultation, analysis and strategy development process that informed the LCW plan and final set of twenty priority actions. The strategy is referred to as 'a transitional strategy for the region' to support the region's growth while limiting the increase in greenhouse gas emissions (WAGA et al., 2014). As commented by the WAGA Executive Officer:

> So the purpose of that strategy of Low Carbon West which is our most important piece of work was to actually take, to change the role, to take it out of being sort of, 'Oh well, what do we do now? What do we do next?' into, 'Well, what actually needs to be done to really make a stepped change to reduce emissions across this region?'
>
> *(WAGA Executive Officer, October 2014)*

After two years researching and developing LCW, a set of twenty actions across four sectors (business and industry, communities, transporting people and freight, urban growth and development) have been developed, summarized in Table 7.2.

The range of proposed actions illustrates the ambitious scope and also the complexity of actors involved. We suggest that, while it is too early to know the extent to which this type of strategy renders legitimate a form of low carbon urbanism that can challenge the status quo of existing urban development and economic agendas, its development seeks to reconfigure the spatial boundaries of low carbon governance at a regional scale, where there is currently no formal tier of government. This is necessary as the capacity to mobilize these types of actions is constrained within local government boundaries yet below the focus of state or federal agendas in most cases. While the strategy can be considered purposive, the range of actions reveals the messy and relational work of low carbon urbanism and low carbon governance and the level of ambition envisaged.

The production of the LCW strategy illustrates the integrative, mediating role that CCAs can play across local government and organizational boundaries, sectors and agencies. In forming a climate change–economic development partnership which "seeks to position the West as a centre for the low carbon economy in Australia" (WAGA Executive Officer, August 2015), LCW explicitly aims to create opportunities to prototype and market test actions with industry, council and community through an iterative process of experimentation, evaluation and learning. Meanwhile, the status of the plan legitimizes both local carbon governance and WAGA in new ways.

TABLE 7.2 Summary of LWC action plan.

Description	Sectors	Actors and intermediaries
Establish a program for smaller industrial energy users to report on energy use and develop resource efficiency plans, and work with large industry energy users to sign up to voluntary agreements for emissions reductions	Business and industry	LeadWest/Open Innovation West (lead), Business Efficiency Network/Australian Industry Group/–Australasian Industrial Ecology Network (linkages), universities (assessment), business awards (state, regional, local), Grow Your Own Business Program/Grow Me the Money (funding)
Facilitate EUAs for energy efficient plant (for both commercial buildings and industrial processing facilities)	Business and industry	Sustainable Melbourne Fund, Department of State Development, Business and Innovation, banks, WAGA & EAGA, local governments
Fund and facilitate industry training for energy efficiency or building tune-ups	Business and industry	Federal Government (funding), Victorian Government (lead/funding), WAGA (lead), LeadWest (network), local governments (funding), businesses (support & mentorship), universities (research)
Create bulk-buy schemes for solar PV panels to be installed in new non-residential buildings	Business and industry; urban growth and development	WAGA (lead), NAGA, EAGA, MEFL, YEF (collaboration), Clean Energy Finance Corp., Sustainable Melbourne Fund (funding), Big Roofs pilot companies – Toll Holdings, GPT, Goodman Property, K-Mart, Toyota, Dulux, Origin, Australand
Implement a White Roofs or Cool Roofs programme	Business and industry; urban growth and development	WAGA (lead), Big Roofs pilot companies, existing facilities with cool roofs, Melbourne Airport (learn from), City of Melbourne (learn from), City of Greater Geelong (collaborator), Wyndham City Council (collaborator)
Provide planning benefits or incentives for high performance building applicants	Urban growth and development	WAGA (lead), local governments, LeadWest

Description	Sectors	Actors and intermediaries
Implement planning scheme requirements for high performance buildings, and advocate for reforms to improve national building standards (such as the Building Code of Australia)	Urban growth and development	WAGA (lead), local governments, LeadWest
Promote sustainability measures to the community through the Urban Sustainability Atlas, particularly for new buildings to identify opportunities for solar PV installations at a proposed location	Urban growth and development; communities	WAGA (lead), Victorian Department of Environment and Primary Industries, local governments (within region) (funding), local governments (outside region) (funding & expansions of tools and resources), businesses (support and links to manufacturers)
Work with real estate agents to make properties more sustainable through retrofits	Communities	WAGA (lead), local governments (implementation of program), real estate companies
Implement planning scheme requirements for high performance buildings; advocate for reforms to improve national building standards (such as the National Construction Code)	Urban growth and development	WAGA (lead), local governments, LeadWest
Create bulk-buy schemes for energy-efficient and renewable technology, such as PV, solar hot-water and efficient lighting to reduce the capital cost for individual consumers	Communities	WAGA (lead), councils (implemention of program), businesses (support and links to manufacturers), GBCA Communities, developers, RMIT & Deakin, Sustainable Suburbs in the West
Advocate for large-scale renewable energy generation in WAGA region (large-scale solar)	Business and industry	WAGA, CEFC, Victorian Government, Sustainable Melbourne Fund, energy companies and generators
Conduct mapping of demand for heating and cooling to identify priority areas for low carbon district heating (the heat map will be an enabler for investment in low carbon district generation)	Business and industry	Federal Government (funding), Victorian Government, energy companies and distributors (funding), WAGA (lead), local governments (funding), businesses (support), Dandenong Precinct Energy Project (collaborator)
Establish a waste-to-energy facility in the region	Business and industry	Regional waste forums, WAGA, Melbourne Water Western Treatment Plant, Ballarat West Employment Zone

Description	Sectors	Actors and intermediaries
Run recycling promotion and education programs, to consolidate and build on current programs	Urban growth and development; communities	Councils (lead), WAGA (collaborator/support), existing community groups, existing landfills or resource recovery facilities within the region
Implement organic waste diversion and compost distribution back to households	Urban growth and development; communities	Councils (lead, implementation), WAGA & MWRRG (collaborator/support), Existing community groups, Existing landfills or resource recovery facilities within the region
An action to establish a regional network of freight businesses and implement a reporting scheme has the potential to significantly reduce emissions associated with the sector	Business and industry; freight, transport and movement	EPA, LeadWest, freight business leaders, WAGA (minor partner),
Continue to promote the West as an ideal site for a freight intermodal transfer station (the centre reduces the demand for heavy trucks on road networks in Melbourne and shifts a degree of interstate freight mode share from trucks to rail, which is less emissions intensive)	Business and industry; freight, transport and movement	MPV, LeadWest/Western Melbourne RDA, WAGA, councils
Establishing (or extending) a formal car-share scheme in the region. (e.g. Flexicar)	Freight, transport and movement; communities	Private sector providers, local government and transport groups (advocacy)
A network and/or apps to encourage residents in the West to share cars for common trips (e.g. to and from work)	Freight, transport and movement, Communities	WAGA, councils (lead by example), businesses and private enterprises, community centres, etc.

Note: EAGA, Eastern Alliance Greenhouse Action; EUAs, Environment Upgrade Agreements; CEFC, Clean Energy Finance Corporation; EPA, Environmental Protection Authority; GBCA, Green Building Council of Australia; MEFL, Moreland Energy Foundation Ltd; MPV, Major Projects Victoria; MWRRG, Metropolitan Waste and Resource Recovery Group; NAGA, Northern Alliance for Greenhouse Action; PV, photovoltaic; RDA, Regional Development Australia; WAGA, Western Alliance for Greenhouse Action; YEF, Yarra Energy Foundation.
Source: adapted from AECOM and Arup, 2014.

Drilling down further into the proposed actions, plans to reframe meanings, materials and knowledge are implicit in the formulation of several of the actions. For example, energy reporting and efficiency plans are standard cleaner production techniques that aim to elevate energy use among smaller companies and to normalize and spread energy management as a practice. Energy Upgrade Agreements both legitimize and legalize new energy services by allowing building upgrades to be funded through specific loans that are paid back through rents. White roof programs seek to intervene in the materiality of the city in deliberate ways but also to spread knowledge of not only the effects of albedo on building efficiency, but also (literally) drawing attention to wider carbon agendas in the process. In a highly industrialized zone that has long hosted 'dirty' industries in Melbourne, ideas of urban and industrial ecology such as waste-to-energy are not new, but through a coalition-building process may be given new meaning, legitimacy and exposure. Examples of ideas to play a role in encouraging the so-called sharing economy through car-sharing apps, etc., similarly express the potential to reframe and legitimize the meanings of low carbon urbanism.

Two shortcomings are apparent when we consider the actual plans of WAGA in this way. First, while plans invariably appear to offer the possibility of reframing meanings, materials or knowledge, they rarely do this explicitly, or in a coordinated way. In other words, they typically offer technology rebates, or new knowledge, or seek to engender new ways of seeing, new norms or new social standards; but they do not do these things in conjunction. Secondly, the plans tend to focus on new business opportunities in the 'green economy' rather than ideas for curbing carbon consumption per se. This makes them inherently dependent upon a narrow interpretation of ecological modernization that, rather than advancing a structural change in development modes, optimistically expects carbon to be displaced by new lower carbon activities. The social and economic structures that underlie (carbon) consumption are of course deeply ingrained and this illustrates one of the constraints of low carbon governance at the local scale in contested political contexts, where overall responsibility for carbon pricing, rationing and budgeting falls outside the remit of the local scale.

6 Discussion and conclusion

Returning to the three key questions set out in the introduction above, this chapter has described how CCAs have emerged. Now we take stock of their governance capacity-building activities and consider the implications for transition activities both locally and more broadly.

Given, the relatively weak role of local governments in Australia, there is an obvious disconnect between the plans and frameworks guiding multilevel climate change policy – which emphasize the importance of local responses – and the capacities of local governments and local actors to deliver. The traditionally weak institutional architecture at the local government scale creates a significant challenge in climate change governance and progressive place governance more broadly. We argue that the emergence and evolving role of regional alliances is reflective of the processes of 'transition' in both the spatial scale of climate responses (from local to regional to statewide) and the institutional processes and practices governing climate change in Victoria. In changing the frame of reference for local governments beyond their individual boundaries to a regional collaboration, the alliances are opening up opportunities to reframe local problems as regional and in the process contribute to processes of structural change.

CCAs as intermediary organizations are enacting at least three processes of mobilizing forms of low carbon urbanism, namely: forms of experimentation seeking to trial upscale demonstration projects across partner councils and statewide (e.g. Environment Upgrade Agreements, solar rooftop initiatives, low carbon economic development strategies); actively driving processes of capacity building and learning across local councils, between local and state governments and across industry and civil society groups; and processes of scaling up and translating low carbon transitioning through multisector and multi-agency projects and strategies. To this extent, they are performing a networked low carbon governance role by mediating, enabling and seeking to legitimize new and reconfigured ways of doing business, urban development and using energy. As organizations that are embedded in particular urban contexts and that have built up relationships and capacities across their regions and between other regions over time, they are practising a form of low carbon urbanism that also intervenes in broader planning (and urban development) projects as an example of progressive place governance. While the processes involved in scaling up systemic transitions is contingent and contested, CCAs are important agents of empowerment elevating local needs and challenges.

We propose that CCAs are themselves rendered legitimate through relational processes (between and within councils, industry and community actors), in their role as experimenters and innovators. There is discernible, ongoing tension between their roles as strategists and activists. Healey (2015) characterizes progressive place governance actors as needing to have both a capacity to shape strategic directions while also being opportunistic. Across each interview, when describing themselves as organizations, the CCA executive officers used words like 'responsive', 'opportunistic', 'strategic', 'nimble' and 'agile'. While this space to experiment was considered critical to their success, the need to secure ongoing funding (i.e. state and federal grants and other funding sources) was critical to moving beyond plans to implementation. Despite this, alliances did not necessarily call for annual funding from state governments, as they believed this might constrain their capacity to be 'nimble' and innovative. They have however made submissions to the state government in recent rounds of consultation, that alliances be authorized to apply directly for government grants rather than through a member of local government and that local governments must be members of an alliance to apply for future 'climate communities' funding (WAGA Executive Officer, August 2015). The Victorian Government's 2017 Climate Change Adaptation Plan includes a specific reference to the state working closely with CCAs recognizing the important role they have been playing for many years with and across local governments.

This recent recognition by state policy-makers strengthens the legitimacy of CCAs as 'cross-border' institutions. In transitioning to a low carbon future in Melbourne, the emergence and continued work of CCAs as critical intermediaries is clear and there is increasing evidence of scaling up efforts to transition to low carbon technologies, infrastructures and ways of doing economic and urban development. These actors do more than addressing local government deficiencies in terms of resourcing and capacity. They are helping to reconfigure and adapt institutional practices necessary to respond more effectively to climate change in their particular regions.

As mediators, they generally lack technological expertise but develop the networks to access it. They are spatially situated with local, 'real' and personal relationships, yet they are reasonably agile in being able to traverse traditional business and local authority boundaries. They are adept networkers and campaigners, seeking to maintain an inside–outside role where they are sufficiently legitimate to be included in mainstream institutional processes, yet also sufficiently distanced to be able to engage with experimental, radical and illegitimate or not-yet-legitimate ideas and stakeholders.

There are three key concerns raised by the progress we have observed to date and the prospects for the future, locally and more broadly. Firstly, there is of course an inherent tension between the temporary and agile nature of the alliances and their ability to maintain a role in transitioning over time. Their legitimacy and their funding is contingent on being both 'outside' and 'inside'. They are embroiled necessarily in power relations within which decision-makers, business and government tend to dominate the visioning and plans towards sustainability, while the weaker segments of society, who can be said to suffer the most from 'unsustainability', are excluded either explicitly or through the discourse, language and mode of engagement adopted. As revealed in a study of the micro-politics of transitioning in the Dutch transport sector (Avelino, 2009) the voices from outside the regime of vested interests in conventional carbon systems are dimmed and excluded over time and whenever such interests are deemed to be at risk. It follows that the fate of the plan rests in coalitions of power gathering across the regime and in the new carbon economy.

Secondly, as indicated in the projects presented in the LCW example, the likelihood of success of the projects in shifting social practices and wider social structures in society is questionable, given that they tend to focus either on technology substitution or information provision. Social practices are held in place by a set of broader social, political and cultural structures and interventions and seeking to shift these must contend with these wider structures. This at least requires attention to social practices and the meanings, materials and competencies that they are constituted by. This, in turn, requires projects that extend beyond behaviour change and so-called rational economic framings of carbon consumption (Strengers et al., 2015) and, in all likelihood, coordinated, multiple interventions that target multiple elements of practices and seek to explicitly end some (high carbon) practices while enabling other (low carbon) ones. This could be addressed by a reframing of the projects, but the fact is they are constrained by the level, resources and limits to governance that are presented by the alliance and by LCW.

This introduces the third concern: the limits of the alliances' scope within contested political-economic structures. As indicated in the LCW example, it is understandable that local governments, with responsibilities for local planning and local economic development, focus on these when formulating low carbon plans and projects. However, in so doing, the wider structures that hold carbon-intensive industries, lifestyles and discourses in place are neither mentioned nor addressed through these projects, leaving open the question of what is the likely fate of these projects as catalysts for wider transition processes. These conclusions point to broader implications for local intermediary organizations and their practices elsewhere, in particular in conditions of contested national policy contexts and weak institutional settings. The prospects for transition are contingent upon local governments working beyond their individual boundaries in regional and inter-actor collaborations in deliberative ways, mindful of ingrained social and economic structures, and focused upon reframing local problems as regional and in the process contributing to processes of structural change.

References

AECOM andArup (2013) *Transition to a Low Carbon Economy Briefing Note: Best Practice Review*. Report prepared for Western Alliance for Greenhouse Action, Victoria.

AECOM and Arup (2014) *Low Carbon West: A Strategy for the Transition to a Low Carbon Economy in the WAGA Region*. Report prepared for Western Alliance for Greenhouse Action, LeadWest and Regional Development Australia (Western Region).

Avelino, F. (2009) Empowerment and the challenge of applying transition management to ongoing projects. *Policy Sciences*, 42(4), 369–390.

Aylett, A. (2015) Institutionalizing the urban governance of climate change adaptation: results of an international survey. *Urban Climate*, 14(1), 4–16.

Bulkeley, H. and Betsill, M. (2003) *Cities and Climate Change: Urban Sustainability and Global Environmental Governance*. London: Routledge.

Bulkeley, H. and Betsill, M. (2005) Rethinking sustainable cities: multilevel governance and the 'urban' politics of climate change. *Environmental Politics*, 14(1), 42–63.

Bulkeley, H., Castán Broto, V. and Edwards, G. (2014) *The Urban Politics of Climate Change: Experimentation and the Governing of Socio-technical Transitions*. London: Routledge.

Bulkeley, H. and Kern, K. (2006) Local government and the governing of climate change in Germany and the UK. *Urban Studies*, 43(12), 2237–2259.

Castán Broto, V. and Bulkeley, H. (2013a) Maintaining climate change experiments: urban political ecology and the everyday reconfiguration of urban infrastructure. *International Journal of Urban and Regional Research*, 37(6), 1934–1948.

Castán Broto, V. and Bulkeley, H. (2013b) A survey of urban climate change experiments in 100 cities. *Global Environmental Change*, 23(1), 92–102.

Fischer, J. and Guy, S. (2009) Re-interpreting regulations: architects as intermediaries for low-carbon buildings. *Urban Studies*, 46(12), 2577–2594.

Geels, F. (2005) The dynamics of transitions in socio-technical systems: a multi-level analysis of the transition pathway from horse-drawn carriages to automobiles (1860–1930). *Technology Analysis & Strategic Management*, 17(4), 445–476.

Guy, S., Marvin, S., Medd, W. and Moss, T. (2011) *Shaping Urban Infrastructures: Intermediaries and the Governance of Socio-technical Networks*. London: Earthscan.

Healey, P. (2015) Civil society enterprise and local economic development. *Planning Theory and Practice*, 16(1), 11–27.

Hodson, M., and Marvin, S. (2010) Can cities shape socio- technical transitions and how would we know if they were? *Research Policy*, 39(4), 477–485.

Hodson, M., Marvin, S. and Bulkeley, H. (2013) The intermediary organisation of low carbon cities: a comparative analysis of transitions in Greater London and Greater Manchester. *Urban Studies*, 50(7), 1403–1422.

Hoffman, M. (2011) *Climate Governance at the Crossroads: Experimenting with a Global Response after Kyoto*. Oxford: Oxford University Press.

Hoppe, T. (2016) Presenting a framework to analyze local climate policy and action in small and medium-sized cities. *Sustainability*, 6(8), 847–889.

Horne, R. and Dalton, T. (2014) Transition to low carbon? An analysis of socio-technical change in housing renovation. *Urban Studies*, 51(16), 3445–3458.

McGuirk, P., Bulkeley, H. and Dowling, R. (2014) Practices, programs and projects of urban carbon governance: perspectives from the Australian city. *Geoforum*, 52, 137–147.

McGuirk, P., Dowling, R., Brennan, C. and Bulkeley, H. (2015) Urban carbon governance experiments: the role of Australian local governments. *Geographical Research*, 53(1), 39–52.

Moloney, S. and Funfgeld, H. (2015) Emergent processes of adaptive capacity building: local government climate change alliances and networks in Melbourne. *Urban Climate*, 14, 30–40.

Moloney, S. and Horne, R., (2015) Low carbon urban transitioning: from local experimentation to urban transformation? *Sustainability*, 7(3), 2437–2453.

Moloney, S., Horne, R. and Fien, J. (2010) Transitioning to low carbon communities – from behaviour change to systemic change: lessons from Australia. *Energy Policy*, 38(12), 7614–7623.

Moore, T., de Haan, F., Horne, R. and Gleeson, B. (eds.) (2017) *Urban Sustainability Transitions – Australian Cases – International Perspectives*. Singapore: Springer.

Moulaert, F. (2010) Social innovation and community development: concepts, theories and challenges. In: Moulaert, F., Martinelli, F., Swyngedouw, E. and Gonzalez, S. (eds.) *Can Neighbourhoods Save the City: Community Development and Social Innovation*. London: Routledge, pp. 4–17.

Parliament of Victoria (2017) *Climate Change Act 2017*. Available from: https://www.climatechange.vic.gov.au/legislation/climate-change-act-2017 (accessed 15 September 2017).

Schlosberg, D. (2015) On the shortcomings of Australian climate policy ahead of COP21. Sydney Environment Institute blog. Available from: http://sydney.edu.au/environment-institute/blog/australian-climate-policy-fails-even-basic-market-ideals/ (accessed 1 April 2016).

Shove, E., Pantzar, M. and Watson, M. (2012) *The Dynamics of Social Practice: Everyday Life and How It Changes*. London: Sage.

Steele, W., Sporne, I., Dale, P., Shearer, S., Singh-Peterson, L., Serrao-Neumann, S., Crick, F., Choy, D.L. and Eslami-Andargoli, L. (2013) Learning from cross-border arrangements to support climate change adaptation in Australia. *Journal of Environmental Planning and Management*, 57(5), 682–703.

Strengers, Y., Moloney, S., Horne, R. and Maller, C. (2015) Beyond behaviour change: practical applications of social practice theory in behaviour change programmes. In Strengers, Y. and Maller, C. (eds.) *Social Practices, Intervention and Sustainability*. London: Routledge, pp. 63–77.

WAGA, Lead West and RDA (2014) *Low Carbon West: A Strategy for the Transition to a Low Carbon Economy in the WAGA Region*. Report prepared by AECOM and Arup for the Western Alliance for Greenhouse Action, Lead West and Regional Development Australia, Melbourne.

8

STRONG LOCAL GOVERNMENT MOVING TO THE MARKET?

The case of low carbon futures in the city of Örebro, Sweden

Mikael Granberg

1 Introduction

Policy and practice at the local level is central in relating global standards and knowledge, national and regional climate change scenarios and policy decisions into particular climate action in a specific context (Bulkeley and Betsill, 2003; Elander et al., 2003; Lundqvist and Biel, 2007; Storbjörk, 2007; Storbjörk, 2010; Castán Broto and Bulkeley, 2012; van den Berg and Coenen, 2012; Romero-Lankao, 2012; Bulkeley et al., 2015). This means that cities, and their local governments, are central to understanding the implementation of international agreements (regimes), national and regional climate change policy. It needs to be stressed, however, that local governments are not just implementers of policy decision taken at higher levels of government. Local governments can, and perhaps have to, be forerunners in climate change policy and practice, as the sources and impacts of climate change are always local, national policy and international negotiations are not always successful, and national governments are not necessarily taking the lead (Gore and Robinson, 2009; Bulkeley et al., 2015; Bulkeley and Betsill, 2003).

Local government action on climate change takes place in a specific local setting. It also takes place in a policy environment characterized by cross-cutting issues and cross pressure from government actors on international, national and regional levels, unfolding public sector reform, continuous policy development, and demands from businesses and citizens (Granberg et al., 2016). Accordingly, why and how cities act on climate change challenges is by no means a straightforward matter (Bulkeley et al., 2015) but, certainly, one that warrants critical research.

This chapter focuses on local government low carbon action within the field of alternative energy production, zooming in on the organizational modes and on intermediary functions and actors in efforts aiming at low carbon transitions (cf. Bulkeley and Betsill, 2013; Hodson et al., 2013). In the Swedish context, low carbon transitions are of course connected to the ecological need to reduce GHG (greenhouse gas) emissions, mitigating future risks and impacts. But they are also increasingly connected to the more economic arguments connected to the concept of a carbon or fossil bubble (Schoenmaker et al., 2015; Rubin, 2015). That is, the idea that any investments made in fossil fuel-based companies are investments that will add to GHC emissions and, perhaps even more important from this perspective, fail to produce any long-term economic profits. According to this perspective, all viable investments need to be directed towards no-carbon solutions, businesses and markets or there will be serious negative economic impacts when the carbon bubble eventually bursts.

The central questions of this chapter are how capacity-building for low carbon transitions evolves at the interface between state and market and what specific role local governments take in this interface. As already indicated above, this perspective is guided by the concept of intermediation through institutional experimentation (cf. Luque-Ayala et al., Chapter 2, this volume; also, Hodson and Marvin, 2009, 2012; Hodson et al., 2013). Intermediaries here are defined as entities that connect, translate and facilitate flows between different parties. The focus of the chapter is, more precisely, on systemic intermediation on a network level involving more than two parties (Hodson and Marvin, 2009). The intermediary role can be divided into facilitating, configuring and brokering (Stewart and Hyysalo, 2008).

Accordingly, the aim of this chapter is to study local government efforts to build low carbon capacity by describing, critically examining and analysing local government climate action. In the case studied here, we can see the local government organization intermediating indirectly by trying to facilitate flows of both experience and capital between public and private actors but also directly as a market actor through a boundary hybrid organization facilitating connections between local government and market actors. A hybrid organization (cf. Koppell, 2006) is an organization that mixes value systems and logics of various spheres such as the state and the market (cf. Erlingsson et al., 2014; Montin, 2016). More precisely, the local government uses a 'green' investment fund, facilitating public–private networking and a municipal company in their efforts to advance a low carbon transition.

The case studied follows the development of local government climate change action over the last decade in the Swedish city of Örebro. The city profiles itself as a forerunner in environmental issues and has formulated ambitious reduction targets. Sweden is often considered a pioneer in environmental governance (Lidskog and Elander, 2012), combining high ambitions at a national level, strong local government and robust policy guided by ecological modernization (Lundqvist, 2000; Zannakis, 2015). It has been stated that if Sweden still holds on to a leading international position in environmental governance it is probably due to activities at the local level (Granberg and Elander, 2007; Uggla and Elander, 2009; Hjerpe et al., 2014). Accordingly, Sweden provides an interesting context for studies of local government climate change action via market mechanisms, given its combination of strong local government with high (national and local) environmental ambitions. Arguably, drawing from this, if Sweden is to be perceived as a 'least likely' case for utilizing market mechanisms due to its strong and resourceful local government (cf. Flyvbjerg, 2006), then local government action in Sweden becomes a critical case worthy of critical inquiry.

In the sections that follow, this chapter elaborates aspects integral to the case studied. First, it briefly presents the Swedish local government system and its development, highlighting the presence of strong local governments and their central role in the Swedish government system. This is followed by a presentation of Swedish policy development within the fields of climate change, energy generation and GHG mitigation, again highlighting the central role given to Swedish local government. The chapter ends with a set of conclusions aimed at an integrated analysis of the Swedish government system, national policy developments and the local component of the case study showing how these three components are, in fact, one integrated multilevel case. It is clear that the local government uses a form of institutional experimentation that mixes value systems and logics of both state and market in its strive to becoming a 'climate smart' city.

2 Strong local government: the Swedish government system and its development

This section gives a short overview of the Swedish government system, with a special focus on the development and role of a strong local government. The section is relevant for understanding the process of change that frames local government climate action in Sweden, as it elaborates on the increasingly mixed ethos of the Swedish government following from the various waves of neo-liberal reforms implemented since the 1970s.

The Swedish government system is an integrative central–local government system, where local governments have substantial constitutional, financial, political and professional resources (Montin and Granberg, 2013; Montin, 2016). Finding ways to develop and expand local governments' institutional capacity has a long history in Sweden. A significant development occurred in the 1950s and 1970s, when the number of local governments was reduced from 2,500 down to 290 (the same as today). The objective of this reform was to enable local governments to assume greater responsibilities and increase efficiency in overall service delivery, including local environmental policy and planning.

Gradually Swedish local government functions and responsibilities have expanded and today they are seen as considerable (Montin and Granberg, 2013). Local government is responsible for a broad range of functions, from childcare to economic development. This means that climate change action within local government intersects (reinforces, complements or conflicts) significantly with other vital responsibilities, including service and infrastructure provision (e.g. local roads, bridges, waste collection and management); planning and development; health and sanitation; welfare services (e.g. elderly care and childcare); building regulation, inspection and controls; facilities management (e.g. aerodromes, cemeteries, parking and ports); culture and recreation (e.g. galleries, libraries and museums); and water provision, sewerage and energy production. Global environmental challenges such as climate change are also on the local government agenda. Similarly, local governments are involved in both the formulation and implementation of strategies for local and regional development (growth policies) and acts in local, national and international contexts (Granberg and Elander, 2007; Granberg, 2008; Montin and Granberg, 2013).

The idea of a strong local government ensuring the efficient production of welfare services, securing the well-being of its inhabitants and facilitating the functions of representative democracy, has been increasingly challenged over the last decades by reforms grounded in a market perspective (Montin, 2016). Since the breakthrough of New Public Management in the 1980s and 1990s, far-reaching reforms have led to an increase in market orientation in Swedish local government (Montin and Granberg, 2013; Montin, 2016). The organizational reforms of local government – primarily focusing on increased economic efficiency – have been based on market approaches as well as models used in the business sector. This development was accentuated during the early years of the twenty-first century through freedom-of-choice reforms and policies focusing on competition, opening up local services to private businesses (as external service providers) while also subjecting internal local government service provision to both internal and external competition (Montin, 2016).

This has entailed a change in the use and perception of government intervention through regulation (Montin, 2016). Regulation concerns governmental intervention in market, or other social processes, to facilitate positive outcomes and control potentially adverse consequences as identified by the processes of representative democracy. Traditionally regulatory measures aiming at carbon control have been commonplace in environmental governance in industrialized countries (While et al., 2010). In the contemporary Swedish debate on the role of local government, conventional regulation is considered a source of negative path dependency, limiting the scope of action for market actors and also undermining competiveness (Montin, 2016). Accordingly, in contemporary Swedish public policy, regulation is used more as a tool for stimulating market responses leading to economic growth and reinforced economic competiveness for both public and private actors rather than regulation as traditional state control (Montin, 2016).

The regulation of public procurement, part of the broad developments described above, has entailed an increased importance of contracts and negotiations influencing the preconditions for political steering and control, widening the gap between politicians and the actual provision of services (Hall et al., 2015; Montin, 2016). Private entrepreneurs increasingly handle the actual provision of services, funded by tax contributions and with local government taking the overarching political and democratic responsibility (at least formally). As part of this, municipal companies (operating under competitive logics) are increasingly playing an important role. In 2014, Swedish local governments used 1,700 municipal companies, employing 48,000 people and with a total worth of approximately 1,875 billion

Swedish crowns (kronor, SEK; about 203 million United States dollars, USD) (Erlingsson et al., 2014).

3 Sweden: mitigation of GHG, energy production and policy development

Since the Swedish government system is heavily decentralized, the national approach to climate mitigation entails a central role for local government. As this section shows, mitigation at the local level is guided by a national climate strategy alongside other national objectives, with the specific objective of reducing climate impacts (Granberg and Elander, 2007; MOE, 2009, 2014). A national climate strategy, the Swedish Climate Strategy, has been in place since 1991, and has been revised repeatedly over the years since then (MOE, 2003). The Swedish government has also formulated sixteen environmental quality objectives, with the objective of 'reduced climate change' considered as the most difficult to achieve (Lidskog and Elander, 2012). Both the strategy and the objective focus primarily on climate mitigation. To achieve the climate strategy and its associated objectives the national government has formulated action plans focusing on energy efficiency and the promotion of renewable energy (MOE, 2014). A section of these action plans focuses on supporting the development and market introduction of new 'climate friendly' technology.

At the regional level, the County Administrative Board (CAB) is responsible for the development and implementation of regional climate action plans in line with the national climate strategy and in collaboration with other stakeholders. The CABs role is to support and coordinate interaction between businesses and local government. One of its areas of work focuses on testing new policy instruments that can facilitate a 'green transition' (MOE, 2014, p. 39). This includes special initiatives supporting the development of wind power technologies (MOE, 2014, p. 40, table 4.1).

Sweden has among the lowest GHG emissions in the EU and in the Organization for Economic Co-operation and Development (OECD, 2011). In the negotiations leading up to the Kyoto protocol, these low national emissions gave Sweden the opportunity to increase its GHG emissions by 4%. Sweden declined this opportunity and went, instead, for substantial reductions of GHG emissions, more in line with its stated national image as "a pioneer in the field of climate policy" (MOE, 2003, p. 16). The Swedish government has used a number of climate policies and instruments, including a carbon tax introduced in 1991 (OECD, 2014) and other energy taxes aiming at phasing out fossil fuels in energy production (MOE, 2014), and in 2018 a climate law came into force. The 2014 national government report to the United Nations Framework Convention on Climate Change showed that the emissions in 2011 were almost 16% lower than in 1990, the baseline year (i.e. from 72.8 to 61.4 million metric tons of CO_2e, carbon dioxide equivalent) (MOE, 2014, p. 28). By 2020 it is estimated that the emissions will be 19% below the 1990 levels, based on existing measures (MOE, 2014, pp. 23–24).

This means that most of the 'low hanging fruits' have already been picked and that further reductions will demand action that drives both technology transitions and institutional transformation.

Swedish action on climate change is guided by the Swedish Climate Strategy (MOE, 2003) and the environmental objectives, and is more concretely formulated in two government bills both entitled *An Integrated Climate and Energy Policy* (MOE, 2009, 2014). The objective is, by 2020, to reduce GHG emissions by 40% (from 1990 levels) and then to reach zero net emissions by 2050 (MOE, 2014, p. 9). This target was sharpened in 2017, aiming at reaching zero net emissions by 2045. In addition, the national government has adopted a proposal for a new climate policy framework, consisting of new climate goals, a climate act and a climate policy council. The Swedish prime minister, Stefan Löfvén, said in 2017 that the climate act "represents an epochal shift for Sweden. Just as we must keep our fiscal house in order, we must also put our house in order regarding climate policy ..." and that one effect will be that Sweden "... will be one of the world's first fossil-free welfare nations" (Government of Sweden, 2017).

Guided by the general direction established by the national environmental policy, Swedish local governments are expected to develop plans and implement local climate action measures. Accordingly, a majority of Swedish local governments have developed and decided on their own local, territorially translated and con-textualized, climate strategies (SEPA, 2010). A central tool in this development is Sweden's extensive local and regional district heating. Almost 90% of multi-dwelling buildings and about 75% of commercial and institutional premises in Sweden are using district heating. Most of the district heating facilities are controlled directly by the local government or through municipal companies (Lundgren et al., 2013). Accordingly, local government control over district heating has been important for introducing renewable energy and shifting out of fossil fuels.

The Swedish mode of energy production is rapidly changing, with an increasing share of biofuels, wind power and nuclear power facilitating a reduction in GHG emissions from energy generation. The share of renewables for energy generation has increased substantially, from 15% in 1970 to 48% in 2010 (MOE, 2014, p. 20). Wind power accounts for only a small part of Swedish energy production (just over 10% in 2015), but its development in recent decades has been quite dramatic, seeing an increase from 52 gigawatt–hours (GWh) in 1986 to 16,300 GWh in 2015 – with a 44.4% increase between 2014 and 2015 alone (SCB, 2016, 2017). In contrast, energy generation from hydropower has not seen any significant changes since the 1970s.

Summing up: Swedish local government development and climate action

As stated in the previous sections, Swedish local government plays a central role in reaching both the ambitious welfare and environmental objectives formulated by the Swedish state. At the same time local government has an action space in the choice of methods used for reaching national objectives and also for additional local government initiatives. Since the 1990s, local government reform and organizational change has shown a clear move to the market, both in general terms as well as in environmental policy (Fudge and Rowe, 2001; Granberg and Elander, 2007; Lidskog and Elander, 2012). Ecological modernization has emerged as the main guiding principle directing the focus towards the economic growth potential of sustainable development and low carbon transitions.

In the Swedish context, new modes of governance are transcending and/or blurring the borders between the state and the market (Granberg and Elander, 2007; Montin, 2016; Bulkeley, 2011). Sweden has among the strongest local government systems in the world and a long tradition of local government responsibility for service delivery and problem-solving within the Swedish welfare state. Despite this background, the frequency of market initiatives and market-inspired approaches is growing rapidly in general (Montin and Granberg, 2013; Erlingsson et al., 2014; Montin, 2016). This chapter's case study is framed by the developments described above, particularly reflecting the specificity of Swedish local government and its more recent search for market-based solutions.

4 The case of Örebro

An environmental forerunner and a 'climate smart' city?

Since the 2011 local elections, the city of Örebro has been governed by a coalition of Social Democrats, the Centre Party and the Christian Democrats (Örebro Municipality, 2016a). Örebro's local government profiles itself as a forerunner in environmental issues and has formulated ambitious reduction targets aiming at a 'climate negative impact' – that is, the local government looks at utilizing carbon capture and long-term carbon storage (carbon sequestration) to counteract inevitable local GHG emissions such as, for instance, those of agriculture (Örebro Municipality, 2015c, 2016a). The local government's sequential reduction targets aim at significant but phased reductions of GHG per inhabitant, as follows: 40% by 2020; 70% by 2030; and 100% by 2050 (baseline 2000, 6.9 metric tons of CO_2e per inhabitant) (Arneson, 2015, p. 14; Örebro Municipality, 2015b, p. 2). In 2012, Örebro's climate impact was 5.8 metric tons of CO_2e per inhabitant, indicating a 16% reduction in the period of 2000–12 and a 10% reduction between 2008 and 2012. The latest evaluation forecasts an estimated reduction of 35% by 2020, a full 5% short of the formulated target of 40%.

Örebro has a strong focus on reducing carbon emissions. In 2015, the idea of the 'carbon bubble' (or 'fossil bubble') became a central concept in a report commissioned by the local government's Finance and Sustainability Department (Arneson, 2015) aimed at exploring the local potential of developing a 'green' economy (cf. Schoenmaker et al., 2015; Rubin, 2015). This idea argues that investments in carbon-based companies and sectors will increase ecological damage and economic vulnerability (Arneson, 2015). Accordingly, public capital investment at a local level has to be handled responsibly, in ways that are good for the environment.

A central issue within the report, beyond its environmental focus, is the position that capital generation is vital in order to make 'green' investments feasible. The argument being that without capital there can be no 'green' investments. Accordingly, it is of central importance for the local government to prioritize climate and environmentally friendly asset/capital management and investments that facilitate both monetary and environmental gain. This is clearly in line with the ecological modernization approach enshrined within national policy and espoused by Prime Minister Stefan Löfvén, as illustrated by his statement discussed earlier in the chapter.

In 2015, the journal *Miljöaktuellt* (Miljöaktuellt 2015) declared Örebro the best Swedish environmental local government. Among the factors mentioned by the journal as motives for this appointment were the local government's use of 'green' bonds (discussed in more detail later in this chapter) and its carbon-free/fossil-free investment policy. As identified by *Miljöaktuellt*, Örebro is focused on using its financial resources coordinated and strategically directed towards facilitating local climate action and low carbon outcomes: "We need to take action on climate change on various levels. Our efforts are more meaningful when we ensure that our financial assets don't work in the opposite direction" (Lena Baastad, the former Social Democratic chair of the City Executive Committee, Örebro, quoted in Collins, 2014).

Efforts to connect financial issues with sustainable development are also evident in local institutional architecture. In 2012, the local government of Örebro established a Finance and Sustainability Department (Klefbom, 2014). The Department is responsible for, among other duties, both the local government's investment strategy and the development of its climate strategy (Örebro Municipality, 2014b, 2016b). The stated aim of this organizational set-up is to integrate economic and financial issues, on the one hand, and sustainability issues on the other. According to the local government, this has enabled the executive branches to track how different departments perform in relation to the objectives and criteria (key performance indicators) of sustainable development. It has also, according to the head of the Finance and Sustainability Department, enabled Örebro to invest in 'green' projects and business, while aiding local government support for 'climate smart' investments from external actors (Hanna Dufva, head of the Finance and Sustainability Department, cited in Gunnarsson, 2015). The aim of the local government is to function as an intermediary connecting local government and market actors, with the aim of facilitating a 'climate smart' development (cf. Hodson et al., 2013).

Within its climate strategy, Örebro also focuses on 'climate smart' local action (Örebro Municipality, 2016b). Central to the 'climate smart' concept is the facilitation of local government collaboration with regional businesses. The aim is to support businesses in the identification of action and foster an exchange of different actors' experiences of climate change action. This, it is argued, would enable climate action that is both efficient in mitigating GHG emissions and reducing climate-related risk while also creating profit for businesses. A central component in this is Örebro Climate Arena, an association/network of local businesses with an interest in spearheading business development in energy and climate issues (Örebro Municipality, 2016b). The emphasis on creating a space where different actors come together to develop local climate action through new business ideas and/or the development of new technologies illustrates the intermediary ambitions of the local government (cf. Stewart and Hyysalo 2008; Kivimaa 2014).

As discussed above, 'climate smart' is a concept embedded both within the local government's climate strategy as well as in its documents concerning investments. It is also a central concept for many other local government documents discussing, promoting and configuring local climate action (Örebro Municipality, 2013, 2014a; Arneson, 2015; Örebro Municipality, 2015b, 2015c, 2016b). Some of the key words connected with the 'climate smart' concept include 'green' and 'climate friendly' economic growth as well as ecological and social sustainability through innovation and strengthened competitiveness. Such documents stress the importance of the integration/unification of these aspects within the local government's policy, planning and practice. It is evident that the local government, in its climate action, has the ambition to be an intermediating public actor facilitating technological innovation and market development (cf. Hodson and Marvin, 2009).

Accordingly, a keen political interest in fostering a connection between local government investments and environmental and renewable energy – including mitigating GHG emissions – is evident in the local government's climate strategy, its investment policy and several other local documents and policies (Örebro Municipality, 2014b, 2016b). The chosen pathway is to advance energy efficiency, to facilitate and support a regional alternative energy market (via 'green' bonds) and to generate investment opportunities for private actors in alternative energy (wind and solar power) through the creation of a municipal energy company. Two of these components, the 'green' bonds and the municipal wind power company, will be discussed in more detail below.

A 'climate smart' local government: investments in alternative energy, energy efficiency and green growth

To facilitate investments, the local government has introduced 'green' bonds (in cooperation with the Swedish bank SEB and the World Bank). The first 'green' bond was issued by the local government in 2014, and had a total value of 750 million SEK (about 87 million USD; Örebro Municipality, 2014a; CICERO, 2014). The second was issued in 2016, with a total value of 500 million SEK, bringing the total value so far to 1,250 million SEK (Örebro Municipality, 2015d). Up to the end of 2015, a total of 546 million SEK of 'green' bonds had been used for investments in projects defined as eligible by the local government; out of this, 249 million SEK had been invested in wind power.

The bonds shall facilitate 'green' investments from both businesses and the public, focusing on both climate change mitigation and adaptation responses. Eligible projects are defined by the local government as:

projects that target (a) mitigation of climate change, including investments in low-carbon and clean technologies, such as energy efficiency and renewable energy programs and projects ('Mitigation Projects') (b) adaptation to climate change, including investments in climate-resilient growth ('Adaptation Projects') or (c) to a smaller extent (max 20%) projects which are related to a sustainable environment rather than directly climate related.

(Örebro Municipality, 2014a, p. 1)

Eligible projects are selected by the Finance Committee (Örebro Municipality, 2014a). The projects deemed eligible for 'green' investments should be integrated into the local government's environment program, its investment program and other environmentally relevant policies. Integration is achieved through linkages between the environment program and the climate plan, the water management plan and the nature conservation plan, the waste management plan and the transport management plan.

Örebro's 'green' fund publishes an annual investor newsletter with information on projects that have received investments, as well as providing more general information on the development of the bonds. This investor newsletter, published on the website of the local government, is provided to the general public for the purpose of facilitating transparency. All the funds from Örebro's 'green' bonds are placed in an earmarked account making such investments easily traceable. The 'green' bonds are also considered as a communications tool facilitating an increased awareness of climate challenges and potential solutions among businesses and investors.

As stated above, the aim is that some of the investments facilitated by the 'green' bonds should drive low carbon transitions through investments in alternative energy solutions giving a profit to investors (the logic for choosing this type of institutional form) and creating an economic surplus for the local government that will contribute towards easing the financial burden of the taxpayers (Örebro Municipality, 2014a). Resources raised by the funds are invested in market actors facilitating 'green' economic development, including the municipal wind power company discussed in more detail below. In that way, the funds utilized by the local government are aimed at promoting the development of an alternative energy market, consisting of semi-public and private businesses. This is in line with the concept of 'endogenous intermediation' as defined by Hodson et al. (2013, pp. 1411–1412).

Using hybrid organizations: alternative energy efforts

Örebro has established, together with the neighbouring city of Kumla, KumBro Utveckling AB, a municipally owned wind power company (Örebro Municipality, 2013, 2016b). The plan is to build fifteen to twenty wind power turbines by 2020, generating 120 GWh a year and meeting the energy demand of about 24,000 regional households (Kumbro Utveckling AB, 2016). This would entail a reduction of 48,960 metric tons of CO_2e per year. To date, as of 2018, 16 turbines are up and running, producing 96 GWh/year. The investment is approximately 565 million SEK (about 63 million USD).

Through the municipal company's wind power, Örebro aims to make local government operations energy self-sufficient. It also aims to support smaller businesses in pursuing 'climate smart' investments while initiating a regional market for wind power-generated energy (Örebro Municipality, 2016c). Citizens can buy shares in the municipal company, and take part in both its future development and the regional alternative energy market. In this way, the local government may be seen as a creating a hybrid boundary organization (i.e. the municipal company; cf. Koppel, 2006) while at the same time it is a configuring, facilitating and expanding regional markets (cf. Stewart and Hyysalo, 2008; Bulkeley et al., 2015).

The government of Örebro stresses that KumBro Utveckling AB must act commercially (Örebro Municipality, 2016c). Yet, at the same time, it acknowledges the fluctuating nature of energy markets, and how this poses a challenge given low prices on energy and high costs in investments. This problematization of economic circumstances has proven accurate, since, as of 2017, KumBro Utveckling AB (and its alternative energy branches) has not yet made profits and has been making a financial loss. The government of Örebro argues that the investment in alternative energy and the efforts to create a viable regional alternative energy market are long-term projects that eventually, and hopefully, will pay off both in terms of GHG emission reductions and financially, but that at this stage of the development the environment is more important than the money (statement from Peter Lilja, CEO of KumBro Utveckling AB, in Wärnelid, 2016).

5 Conclusion

During the last thirty years, Sweden has experienced a transformation in local governance with significant general repercussions for local government discretion, organization and capacity (Montin and Granberg, 2013). The reforms have focused on decentralization, deregulation and citizen participation. Such reforms, with increasing hegemony over later years, have been inspired and guided by catchwords such as privatization, marketization and competiveness. These developments capture the character of such transformation, both for local government in general and for environmental policy and local climate change action in particular (Granberg and Elander, 2007; Montin and Granberg, 2013; Montin, 2016). It is clear that the new modes of governance following the traces of these developments have played a significant part in influencing climate mitigation efforts at both national and local levels. This also includes local government climate action in the city of Örebro.

If we look at the GHG emission reductions, Örebro follows the direction and ambition of national government policy and objectives. The aim is to reduce carbon emissions to zero by the year 2050 with an ultimate target of carbon neutrality. As of 2017, the city has been quite successful already, reducing emissions by 16% between the years 2000 and 2012 and with an estimated reduction of 35% by

2020. A considerable part of the measures instrumental in this strive for carbon reduction take place within local government, focusing on energy efficiency in public buildings and changing fuel sources for local government car fleets and public transport systems, among others.

Other efforts taking place within the local government organization are advanced via the integration of economic and environmental issues, evidenced by the establishment of the Finance and Sustainability Department. The aim is to directly link economic and financial issues with sustainable development, facilitating a 'green' economic transition while advancing the city's low carbon targets. This partly takes place through a form of 'governing by numbers', as it is connected to the mobilization of key economic indicators and sustainable development indicators in the steering and control of local government departments, policy and planning, highlighting 'climate smart' solutions.

It is clear, however, that these efforts are not seen as sufficient by the local government, as this looks for other ways of reaching its GHG emission reduction targets. Örebro chooses to act as an intermediary boundary organization, configuring and facilitating a space that includes opportunities for non-state actors while also stimulating economic growth in its strive towards a low carbon transition (cf. Stewart and Hyysalo, 2008; Hodson et al., 2013). These actions build, at least in part, on the idea that there has to be a direct connection between local public investments, a reduction in GHG emissions (guided by the 'carbon bubble' concept) and the actions of private business actors.

Accordingly, the case of Örebro is an example of a local government functioning as an intermediary that is looking for new modes of governing aiming at breaking away from path dependency in conventional institutionalized local government practices – a feat advanced by experimenting with alternative institutional set-ups (cf. Bulkeley et al., 2015). The rationality is that, in order to increase efficiency and speed up the low carbon transition, new modes of organizing, steering and governing are needed. Accordingly, market-inspired modes of organizing local governance are developed and mobilized with the aim of developing intermediary structures between local government and market actors in ways that increase the efficiency and speed of the low carbon transition (Hodson et al., 2013).

Two main modes are used here: facilitating a market approach by means of 'green' bonds and the creation of a municipal wind power company as a boundary hybrid organization. The local government is using market organizational models and economic incentives to drive an institutional and socio-technical transition towards a low carbon future – one that includes the ambition to facilitate, configure and broker regional and alternative (energy) markets (Stewart and Hyysalo, 2008). The aim is for increased efficiency in emissions reduction, investor profit and an economic surplus for the local government easing the burden on, and risk taken by, taxpayers.

By concrete measures, looking at the broad output (financial and environmental) of the municipal company, these aspirations seem to have been largely unfulfilled (Wärnelid, 2015, 2016) as the development of a regional market for alternative energy is slow and investments are much lower than expected. The return on investments is not sufficient and this reduces the attractiveness of investments based purely on economic calculations (the market logic). This means that this is, at the moment, a bad business deal for taxpayers, as the local government is subsidizing the continuation of business for the municipal company. The counter argument from the local government has been that the environment is more important than the money (Wärnelid, 2016). The problem with this argument is that the reduction in carbon is also quite modest at this stage in time.

Would it, however, be possible to say that the judgment on the ability of the local government of Örebro to reach its objectives – of securing both a technological transition with reduced GHG emissions and a transformation of the local government institutional setting – through market-inspired approaches is premature? The local government's investment policy builds on the concept of a 'carbon bubble'. This idea also guides the introduction of the 'green' bonds and development of the municipal wind power company. In line with the 'carbon bubble' concept, any investment in fossil fuel-based companies or high carbon economies are investments that will slow down technological development, add to GHG emission and forfeit economic returns on investments when the bubble eventually bursts and the carbon economy collapses. Since this situation has not yet materialized, the long-term efficiency of the approach in economic, technological and ecological terms is difficult to judge at this stage in time.

Let's end this chapter by widening our gaze and more theoretically and philosophically reflect on the findings of this case study for issues of democratic accountability in local government and the potential ability of public institutions to control and steer low carbon transitions as intermediaries relying on market models. Institutional experimentation has the potential to be both a resource and barrier to climate action (cf. Bulkeley et al., 2015), something that depends on the specific perspective used as much as the value system utilized. However, organizations that mix value systems and logics of various sectors, such as the state and the market (Koppell, 2006), are, potentially, less susceptible to political and democratic control largely due to the fact that they behave as external entities rather than as extensions of the public politics, democratic processes and administration of local government (cf. Koppell, 2006, 2010; Erlingsson et al., 2014). In line with that argument, the integration of public and market logics within the local governance of climate change runs the risk of undermining political and democratic control of local climate action, by letting market logics frame, and even dominate, local government action (Montin, 2016). The case study presented in this chapter does not fully support such a critical conclusion, but raises important questions on the tensions between efficiency and democratic accountability that warrant future research on the intermediary role of local government in low carbon transitions.

References

Arneson, H. (2015) *Grön ekonomi – Omställning till en klimatvänlig, resurseffektiv och socialt inkluderande ekonomi genom kapitalförvaltning: Bakgrundsrapport till ny klimatstrategi för Örebro kommun.* Örebro: Finance and Sustainability Department, Örebro Municipality.

Bulkeley, H. (2011) Cities and subnational governments. In: Dryzek, J.S., Noorgard, R.B. and Scholsberg, D. (eds.) *The Oxford Handbook of Climate and Society*. Oxford: Oxford University Press, pp. 464–478.

Bulkeley, H. and Betsill, M.M. (2003) *Cities and Climate Change: Urban Sustainability and Global Environmental Governance*. New York: Routledge.

Bulkeley, H. and Betsill, M.M. (2013) Revisiting the urban politics of climate change. *Environmental Politics*, 22(1), 136–154.

Bulkeley, H., Castán Broto, V. and Edwards, G.A.S. (2015) *An Urban Politics of Climate Change: Experimentation and the Governing of Socio-technical Transitions*. New York: Routledge.

Castán Broto, V. and Bulkeley, H. (2012) A survey of urban climate change experiments in 100 cities. *Global Environmental Change*, 23(1), 92–102.

CICERO (Center for International Climate and Environmental Research) (2014) *'Second Opinion' on Örebro kommun's Green Bond Framework*. Oslo: CICERO.

Collins, J. (2014) Ever more cities divesting from fossil fuels. *Deustche Welle*. Available at: http://www.dw.de/ever-more-cities-divesting-from-fossil-fuels/a-18033130?maca=en-rss-en-all-1573-rdf (accessed 6 November 2014).

Elander, I., Granberg, M., Gustavsson, E. and Montin, S. (2003) *Climate Change, Mitigation and Adaptation: The Local Arena*. Örebro: Centre for Urban and Regional Studies.

Erlingsson, G.Ó., Fogelgren, M., Olsson, F., Thomasson, A. and Öhrvall, R. (2014) *Hur styrs och granskas kommunala bolag? Erfarenheter och lärdomar från Norrköpings kommun*. Report 2014:6, Linköping: Center for Communal Strategic Studies, Linköping University.

Flyvbjerg, B. (2006) Five misunderstandings about case-study research. *Qualitative Inquiry*, 12(2), 219–245.

Fudge, C. and Rowe, J. (2001) Ecological modernisation as a framework for sustainable development: a case study in Sweden. *Environment and Planning A*, 33(9), 1527–1546.

Gore, C. and Robinson, P. (2009) Local government response to climate change: our last, best hope? In: Selin, H. and VanDeveer, S.D. (eds.) *Changing Climate in North American Politics: Institutions, Policymaking and Multilevel Governance*. Cambridge, MA: MIT Press, pp. 138–158.

Government of Sweden (2017) Government proposes historic climate reform for Sweden. Stockholm: Government Offices of Sweden. Available at: http://www.government.se/press-releases/2017/02/government-proposes-historic-climate-reform-for-sweden/ (accessed 3 April 2017).

Granberg, M. (2008) Local governance in 'Swedish': globalisation, local welfare government and beyond. *Local Government Studies*, 34(3), 363–377.

Granberg, M. and Elander, I. (2007) Local governance and climate change: reflections on the Swedish experience. *Local Environment*, 12(5), 537–548.

Granberg, M., Nyberg, L. and Modh, L.-E. (2016) Understanding the local policy context of risk management: Competitiveness and adaptation to climate risks in the city of Karlstad, Sweden. *Risk Management*, 18(1), 26–46.

Gunnarsson, J. (2015) Hemligheten bakom Örebros förstaplats. *Miljöaktuellt*, 25 June. Available at: http://www.aktuellhallbarhet.se/hemligheten-bakom-orebro-forstaplats/ (accessed 15 July 2017).

Hall, P., Löfgren, K. and Peters, G. (2015) Greening the street-level procurer: challenges in the strongly decentralized Swedish system. *Journal of Consumer Policy*, 39, 467–483.

Hjerpe, M., Storbjörk, S. and Alberth, J. (2014) 'There is nothing political in it': triggers of local political leaders engagement in climate adaptation. *Local Environment*, 20(8), 855–873.

Hodson, M. and Marvin, S. (2009) Cities intermediating technological transitions: understanding visions, intermediation and consequences. *Technological Analysis & Strategic Management*, 21(4), 515–534.

Hodson, M. and Marvin, S. (2012) Mediating low-carbon urban transitions? Forms of organization, knowledge and action. *European Planning Studies*, 20(3), 421–439.

Hodson, M., Marvin, S. and Bulkeley, H. (2013) The intermediary organisation of low carbon cities: a comparative analysis of transitions in Greater London and Greater Manchester. *Urban Studies*, 50(7), 1403–1422.

Kivimaa, P. (2014) Governmnet-affiliated intermediary organisations as actors in system-level transitions. *Research Policy*, 43(8), 1370–1380.

Klefbom, E. (2014) Så integrerar Örebro hållbarhet och ekonomistyrning. *Miljöaktuellt*, 17 October. Available at: http://www.aktuellhallbarhet.se/sa-integrerar-orebro-hallbarhet-och-ekonomistyrning/ (accessed 15 July 2017).

Koppell, J.G.S. (2006) *The Politics of Quasi-government: Hybrid Organizations and the Dynamics of Bureaucratic Control.* Cambridge: Cambridge University Press.

Koppell, J.G.S. (2010) Administration without borders. *Public Administration Review*, 70 (suppl. 1), 46–55.

KumbroUtvecklingAB (2016) Självförsörjande på förnybar el år 2020. Örebro: Kumbro Utveckling AB. Available at: http://kumbro.se/kumbro-vind/ (accessed 14 October 2016).

Lidskog, R. and Elander, I. (2012) Ecological modernisation in practice? The case of sustainable development in Sweden. *Journal of Environmental Policy and Planning*, 14(4), 411–427.

Lundgren, T., Stage, J. and Tangerås, T. (in collaboration with Carlén, B.) (2013) *Energimarknaden, ägandet och klimatet.* Stockholm: SNS Förlag.

Lundqvist, L.J. (2000) Capacity-building or social construction? Explaining Sweden's shift towards ecological modernisation. *Geoforum*, 31(1), 21–32.

Lundqvist, L.J. and Biel, A. (2007) From Kyoto to the town hall: transforming national strategies into local and individual action. In: Lundqvist, L.J. and Biel, A. (eds.) *From Kyoto to the Town Hall: Making International and National Climate Policy Work at the Local Level.* London: Earthscan, pp. 1–12.

Miljöaktuellt (2015) De är Sveriges bästa miljökommuner. *Miljöaktuellt*, 26 June. Available at: http://www.aktuellhallbarhet.se/de-ar-sveriges-basta-miljokommuner/ (accessed 14 October 2016).

MOE (Ministry of the Environment) (2003) *The Swedish Climate Strategy.* Stockholm: MOE. Available at: http://www.government.se/information-material/2003/01/the-swedish-climate-strategy-a-summary/ (accessed 15 July 2017).

MOE (Ministry of the Environment) (2009) *En sammanhållen klimat- och energipolitik – Klimat.* Prop. 2008/09:162. Stockholm: MOE. Available at: http://www.regeringen.se/contenta ssets/cf41d449d2a047049d7a34f0e23539ee/en-sammanhallen-klimat–och-energipoli tik—klimat-prop.-200809162 (accessed 15 July 2017).

MOE (Ministry of the Environment) (2014) *Sweden's Sixth National Communication on Climate Change under the United Nations Framework Convention on Climate Change.* Ds 2014:11. Stockholm: MOE. Available at: http://www.government.se/contentassets/94d274fef8ef470a 9b6901421b50d1d1/swedens-sixth-national-communication-on-climate-change—under-the-united-nations-framework-convention-on-climate-change-ds-201411 (accessed 15 July 2017).

Montin, S. (2016) En moderniserad kommunallag i spänningsfältet mellan politik och marknad. *Statsvetenskaplig tidskrift*, 118(3), 339–357.

Montin, S. and Granberg, M. (2013) *Moderna kommuner.* 4th ed. Stockholm: Liber.

OECD (Organization for Economic Co-operation and Development) (2011) *OECD Economic Surveys Sweden.* Paris: OECD Publishing.

OECD (Organization for Economic Co-operation and Development) (2014) *Environmental Performance Reviews Sweden 2014.* Paris: OECD Publishing.

Örebro Municipality (2013) *Hållbar utveckling i Örebro kommun: Temarapport Klimat 2013.* Dnr, Ks 537/2013. Örebro: Örebro Municipality. Available at: http://www.orebro.se/download/18.1d8f9a39155628f738416835/1467966336307/Hållbar+utveckling+i+Örebro+ kommun+-+temarapport+2013.pdf (accessed 15 July 2017).

Örebro Municipality (2014a) Örebro kommun – green bond framework. Örebro: Örebro Municipality. Available at: https://www.orebro.se/download/18.2bea29ad1590bf258c 52839/1484206883932/Örebro+Kommun+Green+Bond+Framework.pdf (accessed 15 July 2017).

Örebro Municipality (2014b) *Placeringsriktlinje för Örebro kommun.* Örebro: Örebro Municipality. Available at: https://www.orebro.se/download/18.1d8f9a39155628f7384168bd/ 1467966353022/Placeringsriktlinje+för+Örebro+kommun.pdf (accessed 15 July 2017).

Örebro Municipality (2015a) *Miljöprogram Örebro kommun: Örebro kommuns miljömål och prioriteringar i miljömålsarbetet.* (Updated 2 February 2017.) Örebro: Örebro Municipality. Available at: https://www.orebro.se/download/18.1d8f9a39155628f73841694d/148656 1008915/Miljöprogram+Örebro+kommun.pdf (accessed 15 July 2017).

Örebro Municipality (2015b) *Underlagsrapport till Klimatstrategi för Örebrokommun.* Örebro: Örebro Municipality.

Örebro Municipality (2015c) *Klimatstrategi för Örebro kommun: Mål och delmål för 2020 och 2030* [draft version]. (Updated 10 June 2016.) Örebro: Örebro Municipality. Available at: https://www.orebro.se/download/18.1d8f9a39155628f738416949/1485339455953/ Klimatstrategi+Örebro+kommun+underlagsrapport.pdf (accessed 15 July 2017).

Örebro Municipality (2015d) Gröna obligationer. Örebro: Örebro Municipality. Available at: http://www.orebro.se/fordjupning/fordjupning/sa-arbetar-vi-med/klimat—miljoarbete/ grona-obligationer.html (accessed 13 July 2017).

Örebro Municipality (2016a) Kommunfullmäktige. Örebro: Örebro Municipality. Available at: http://www.orebro.se/kommun–politik/politik–beslut/kommunfullmaktige.html (Accessed 14 October 2016).

Örebro Municipality (2016b) *Klimatstrategi för Örebrokommun: Mål och delmål för 2020 och 2030.* Örebro: Örebro Municipality. Available at: https://www.orebro.se/download/18. 1d8f9a39155628f73841694a/1485339455931/Klimatstrategi+Örebro+kommun.pdf (accessed 15 July 2017).

Örebro Municipality (2016c) *Underlagsrapport till Klimatstrategi för Örebro kommun.* Örebro: Örebro Municipality.

Romero-Lankao, P. (2012) Governing carbon and climate in the cities: an overview of policy and planning challenges and options. *European Planning Studies,* 20(1), 7–26.

Rubin, J. (2015) *The Carbon Bubble: What Happens to Us When it Bursts.* Toronto: Penguin Random House.

SCB (Statistiska centralbyrån/Statistics Sweden) (2016) El-, gas- och fjärrvärmeförsörjningen 2015, definitiva uppgifter. Statistiska meddelande EN 11 SM 1601. Örebro: SCB.

SCB (Statistiska centralbyrån/Statistics Sweden) (2017) El från vindkraft ökar I Sverige. Örebro: SCB. Available at: http://www.scb.se/hitta-statistik/sverige-i-siffror/miljo/ (accessed 3 April 2017).

Schoenmaker, D., van Tilburg, R. and Wijfells, H. (2015) What role for financial supervisors in addressing systemic environmental risks? Sustainable Finance Lab working paper. Utrecht: University of Utrecht School of Economics.

SEPA (Swedish Environmental Protection Agency) (2010) *Gör arbetet med klimatstrategier någon skillnad? En utvärdering av lokalt klimatstrategiarbete.* Stockholm: SEPA.

Stewart, J. and Hyysalo, S. (2008) Intermediaries, users and social learning in technological innovation. *International Journal of Innovation Management,* 12(3), 295–325.

Storbjörk, S. (2007) Governing climate adaptation in the local arena: challenges of risk management and planning in Sweden. *Local Environment*, 12(5), 457–469.

Storbjörk, S. (2010) 'It takes more to get a ship to change course': barriers for organizational learning and local climate adaptation in Sweden. *Journal of Environmental Policy and Planning*, 12(3), 235–254.

Uggla, Y. and Elander, I. (2009) Kommunerna och klimatet: tendenser, möjligheter och problem. In: Uggla, Y. and Elander, I. (eds.) *Global uppvärmning och lokal politik*. Stockholm: Santérus Academic Press, pp. 129–136.

van den Berg, M. and Coenen, F. (2012) Integrating climate change adaptation into Dutch local policies and the role of contextual factors. *Local Environment*, 17(4), 441–460.

Wärnelid, G. (2015), Förslag: 56 friska miljoner till vindkraft. *Nerikes Allehanda*, 24 August. Available at: http://www.na.se/orebro-lan/forslag-56-friska-miljoner-till-vindkraft (accessed 15 July 2017).

Wärnelid, G. (2016) Miljonvinster för kommunens bolag – vindkraften enda plumpen. *Nerikes Allehanda*, 25 April. Available at: http://www.na.se/orebro-lan/orebro/miljonvin ster-for-kommunens-bolag-vindkraften-enda-plumpen (accessed 15 July 2017).

While, A., Jonas, A.E.G. and Gibbs, D. (2010) From sustainable development to carbon control: eco-state restructuring and the politics of urban and regional development. *Transactions of the British Institute of Geographers*, 35(1), 76–93.

Zannakis, M. (2015) The blending of discourses in Sweden's "urge to go ahead" in climate politics. *International Environmental Agreements: Politics, Law and Economics*, 15(2), 217–236.

9

EXAMINING URBAN AFRICA'S LOW CARBON AND ENERGY TRANSITION PATHWAYS

Jonathan Silver and Simon Marvin

1 Introduction

This chapter develops a critical response to how we might undertake analysis of low carbon transitions in sub-Saharan African towns and cities. In doing so the chapter seeks to reflect upon the analytical value and limits of urban transitions analysis (UTA) in understanding how Africa's 'urban revolution' (Parnell and Pieterse, 2014) will intersect with low carbon imperatives in the shadow of the growing climate crisis. We explore how these potentials and deficits can be addressed by examining promising debates across urban studies and how they can help sensitize UTA to the geographies of low carbon transition across urban Africa. While scholarship on urban low carbon transitions has grown substantially over recent years much of this work has taken place in the global North. This chapter offers a provocation that seeks to shift these knowledges and debates into new contexts, through a focus on the relations between an ongoing energy transition, the broader low carbon transformation of cities and explanatory frameworks for such transformation.

It is important to think about low carbon energy transitions in Africa through an urban studies lens for two key reasons. First, the importance of energy in the rapid urbanization of Africa. There are two dimensions to this engagement – the relations between energy and the development agenda in which urban regions are associated with very low rates of connection to formalized energy infrastructure and the reliance on informal provision that this engenders – together with the critical importance of energy in supporting wider urban economic growth imperatives. These challenges, tensions and competing priorities are often most keenly experienced in urban regions. Second, the issue of how urban governance of low carbon interrelates with the various (and multi-scaled) geographies of the energy system split into generation, transmission and distribution with complex ownerships from privatized (sometime liberalized) to state control (Graham and Marvin, 2001). The central issue here is to what extent is it possible for urban governance actors to develop decarbonization-related priorities and interventions when they are often not part of the formalized, socio-political organization of the wider energy system. Additionally, these shifting geographies of a low carbon transition prompt the need to consider how urban governance actors seek to position the multiple spaces of the energy system to implement socially and environmentally just forms of experimentation and transformation.

We address these issues through a critical engagement with UTA in which we seek to identify the potential value of the approach to the African urban context by identifying its limitations but then constructively seeking to expand and reformulate the approach, building on recent work that has brought together a multilevel perspective and urban governance. We then use insights drawn from across urban studies to suggest four contributions to an expanded UTA of low carbon that is sensitive to the geographic context of urban Africa. First, we draw on debates in postcolonial urban studies to suggest ways to account for wider urbanization trends and particularly the critical role of informal energy networks, multiple, overlapping and unintegrated systems in meeting energy requirements. Here, we argue for the need for UTA of low carbon to engage with a fragmented, dispersed and often unregulated set of infrastructures outside the formal system of (urban) energy governance (Figure 9.1). Second, we mobilize debates concerning geography and

FIGURE 9.1 Solar panel in Enkanini informal settlement as part of the iShack project, Stellenbosch, South Africa.
Source: the authors.

scale within UTA to argue for the need to deal with the specificity of how towns and cities relate to the management of low carbon transition. This is generated through the institutional context in which many municipalities lack both the formal capacity and the resources to choose to become active in energy planning, implementation and technology development. Third, we bring attention via debates on urban (infrastructure) governance in the global South to the critical role of wider social interests – publics, the urban poor, civil societies, social movements and local communities – and what sorts of agency can be developed among these urban intermediaries in relation to both low carbon and energy futures. Finally, we use urban political ecology to draw focus on how power, politics and existing urban inequalities shape the way in which decarbonization is envisioned, contested, circulates and is materialized by a range of urban actors.

2 Powering Africa's urbanization

Despite sub-Saharan Africa having the lowest urbanization levels of all global regions, rapid urban growth on the continent is expected to see over 700 million urban dwellers by 2030 (UN–Habitat, 2010) and 1.2 billion by 2050 (UN–DESA, 2009). Such significant expansion of towns and cities is shaping what Parnell and Pieterse (2014) term 'Africa's Urban Revolution' – a dramatic rendering of the complex, shifting geographies of the continent. This rapid urbanization will account for nearly all population growth in Africa over the coming decades, creating multiple infrastructural imperatives for governments, cities and communities alike. Balancing the necessities of economic growth, sustainability, the plethora of national policy objectives and the needs of the 40 to 60% of urban dwellers living in poverty (Toulmin, 2009) we suggest that urban infrastructures will be at the centre of development efforts (Khennas, 2012). As such it is clear that these socio-technical systems will be placed under a series of growing everyday pressures, together with playing a central role in mediating the multiple and open futures of this 'urban revolution' (Parnell and Pieterse, 2014).

One of the most important infrastructures supporting this second wave of urbanization is the provision of energy services to power the burgeoning towns and cities of the continent (Madlener and Sunak, 2011). Energy networks are vital to sustaining the everyday urban life of sub-Saharan Africa through a series of flows that provide electricity, charcoal, firewood, gas and so forth to homes, businesses and the public sector. Energy is pivotal to the African urban transition. As the recent UN–Habitat (2014, p. 41) *State of African Cities* report makes clear, "[t]he growth of Africa's energy sector is a prerequisite for sustained expansion in all others." Yet here exists a conundrum, as McDonald (2009, p. xv) argues, "Africa is the most under-supplied region in the world when it comes to electricity, but its economies are utterly dependent on it." In 2012 the UN declared the 'International Year for Sustainable Energy' highlighting global concern for achieving a series of low carbon policy initiatives and placing energy at the centre of debates about the delivery of the Millennium Development Goals (UN–Energy, 2005), the subsequent, post-2015 Sustainable Development Goals and of course various climate change agreements such as the Paris Accord. A focus on energy across these debates reinforces how the governing of flows and circulations becomes a prerequisite mediating all manner of development indicators within and beyond Africa from poverty through to heath, education, gender equality and civic participation (Brew-Hammond, 2010).

As numerous policy publications make explicit (ICA, 2009; UN–Habitat, 2014) the investment deficit for infrastructure including energy is significant, estimated at 360 billion USD (United States dollars) for Nigeria alone (Simone, 2010b), producing widespread breakdown and disruption (Khennas, 2012; Silver, 2015) and curtailing basic rights in accessing technologies and essential services (UN–Habitat, 2014). In Malawi, Burundi and Liberia less than 10% of the population are able to access modern electricity networks (IEA, 2014). But attention is increasingly being directed to new investment such as the Grand Inga Dams project in western Democratic Republic of the Congo that aims to generate 40,000 megawatts of power or the South African parastatal Eskom's (2012) ambitious Electrification Roadmap for South Africa, Africa and Developing Countries, which aims to connect 500 million people to networked energy services in over 50 countries across the continent. Eskom shows the geopolitical aspirations of South Africa in the energy arena of the continent and the new markets that seem likely to open (McDonald, 2009). This is reflected in the increasing amounts of international capital investment and attention across both low carbon and traditional energy sectors (Power et al., 2016). At a national scale countries such as Ghana are broadening their ambitions from electrification to engage with wider issues of carbon reduction, climate change mitigation and renewable energy (Ghana Energy Commission, 2011).

What these and the plethora of initiatives, projects, financing and emerging institutional arrangements show is that energy has, over the last few decades, become an important concern to multiscalar governance actors and the development aspirations of the continent (Kebede et al., 2010; Sokona et al., 2012). And urban contexts are becoming increasingly important sites in these processes as global goals around low carbon are translated at multiple scales and across geographically stretched networks.

Yet there are important differences in the ways that urban energy demands, infrastructure development and future low carbon plans across sub-Saharan Africa are (re)shaped by the region's distinct and multiple geographies. These are, of course, also very different to the development of modern urban energy services in the global North (Rutherford and Coutard, 2014). The rising numbers of urban poor form the central demographic in the rapid growth of African cities, estimated at over 100 million in the next decade alone (Parnell and Walawege, 2011). These populations demand access, electrification and affordable service provision from across the formal and informal infrastructure conditions that make up energy networks in the low-income neighbourhoods that constitute significant parts of many cities. These demands from the urban poor sit alongside a growing middle class, implicated in intensifying usage, requiring increased generation to facilitate emerging consumption patterns (Karekezi and Majoro, 2002; Silver, 2015). As Rutherford and Coutard (2014, p. 1356) helpfully suggest, "[u]rban energy transition in the South thus clearly means something very different from the North, combining issues around governance, access to finance, trade and supply chains with everyday concerns of, amongst other things, very low basic household incomes, availability of cooking fuel and indoor air pollution."

There are also global imperatives and issues that apply in sub-Saharan Africa as elsewhere that will shape a series of differentiated low carbon transition pathways (Bulkeley et al., 2010). For instance the need to reduce greenhouse gas emissions globally is also predicating new ways to understand Africa's urban energy transition as a low carbon transition. As Bridge et al. (2013, p. 331) comment, "Ensuring the availability and accessibility of energy services in a carbon-constrained world will require developing new ways – and new geographies – of producing, living, and working with energy." We suggest that the low carbon, urban energy transitions that respond to these global–local energy dynamics and imperatives generate a series of strategic pressures on how innovation across infrastructure space is organized that require more detailed attention.

3 Urban transitions analysis

There is emerging but convincing evidence then that urban governance actors are under pressure to strategically respond to the regional energy geographies outlined above through developing managed, systemic low carbon change in the socio-technical organization of key aspects of their energy networks (Swilling and Annecke, 2012). In this section we will examine how UTA (urban transitions analysis), developed predominantly through work across global North cities, can help us to understand the unfolding socio-spatial dynamics of low carbon (Bulkeley et al., 2010).

UTA has emerged from a constructive and critical engagement between the multilevel perspective and debates focusing on governance within urban studies. The multilevel perspective on transitions provides a framework for understanding how infrastructure systems are organized, (temporarily) stabilized and how technological change can be incremental through to radically transformative (Geels, 2002; Geels and Schot, 2007; Smith et al., 2005; Verbong and Geels, 2010). In doing so it seeks to understand the institutional and technological conditions through which a socio-technical regime can be reshaped through internal pressures, wider socio-technical landscape pressures, and through the development of protected niches such as government subsidy for solar implementation. While the multilevel perspective approach is useful in understanding the distributed nature of technological transitions our critical concern is that institutions, social interests and knowledge are often assumed to be 'out there'. Place is implicit within the division of landscape, regime and niche. Only within the niche is the notion of place more explicit as some sort of bounded, experimental and local context while the landscape and regime are assumed to be global, international or national scale in nature. Spatial scale frequently remains implicit or underdeveloped in the multilevel perspective.

The consequence of this is that we are often unclear about where transitions, including low carbon, take place and, given the mutual shaping of system and social context, the spaces and places where transitions take place (Bridge et al., 2013; Bulkeley et al., 2010; Hodson and Marvin, 2010, 2012). There is a developing body of work that addresses the urban and regional geographies of transition and questions of scale and space, including the low carbon dimensions of such change (Hodson and Marvin, 2010; Bulkeley et al., 2010; Coenen and Truffer, 2012; Coenen et al., 2012; Geels, 2010; Raven et al., 2012a; Rohracher and Spath, 2014; Spath and Rohracher, 2010, 2012; Truffer and Coenen, 2012). We are particular interested in the two critical insights this work offers in thinking about a richer and more geographically informed conception of UTA.

First, UTA requires an appreciation of 'multilevel governance' and the politics of scale in understanding attempts to shape low carbon transitions in particular urban contexts. Key to this is the development of renewed approaches to these dynamics, that draw on learning over the last decade across diverse geographical processes of low carbon transformation, such as that offered in the framing of this volume and broadly understood as a stepping stone towards a second generation of studies on urban transition (see Chapter 2). Incorporating these concerns permits us to analyse the entangled relations and interactions of governance at the different scales that inform potential low carbon transition. This is particularly significant in a period of accelerating and unstable globalization, shifting political-economic connections, networks, markets and flows of investment. Here we draw attention to the changing role of the state concerning new forms of decarbonized infrastructure and issues of multilevel governance. Critically, this requires, "an appreciation of the complex geometry of power and the political and cultural struggles through which societies assume their regional shape" (MacLeod and Jones, 2001, p. 670). A focus on the 'endogenous' city and region and 'creating the conditions' for low carbon transitions often ignores what drives urban–regional economies of infrastructure and in doing so underplays the differential economic, ecological and political positions of places and the wider role of the nation-state in devolving responsibility (but not power and resources) for technology and innovation strategies (Ward and Jonas, 2004). Critically questioning these relations between scales allows us to conceive of urban regions not merely as sites for receiving national or international low carbon transition initiatives but also potentially as contexts for the development of more purposive transitions that may address local rather than national priorities (Bridge et al., 2013; Bulkeley et al., 2010; Hodson and Marvin, 2010).

Second, urban governance actors are then clearly constructed in an unfolding and structured set of social, political and economic relations of low carbon. Importantly new renewable technology and innovation is both a product of, and produces political pressures for, institutional change. Seeing the urban as merely responding to these decarbonization imperatives ignores specific relations informing urban contexts and attempts to shape infrastructure space through multiscalar governance. This highlights the importance of seeing urban infrastructure in relation to technological transitions not only through the lens of 'endogenous' institutional interrelationships, but also in terms of the influence of and relations with the nation-state and global institutions. Urban governance can actively and strategically work both internally and externally in developing the resources, networks and relationships to actively shape low carbon transitions. Considering how the capacity to act is organized in constituting and realizing these shifts in energy production, distribution and operation in such a fragmented context is therefore an important issue. Understanding the dynamics of how such spaces of intervention are created and maintained is therefore fundamental for understanding the various attempts to develop and translate visions of an energy transition into the future of cities (Bulkeley et al., 2010; Coenen and Truffer, 2012; Geels, 2010; Hodson and Marvin, 2010; Raven et al., 2012b; Rohracher and Spath, 2014; Spath and Rohracher, 2010, 2012).

Critically then work on UTA has started to provide a conceptual outline through which different urban contexts mediate variable capacity to envision and enact low carbon transitions. Key to this is understanding the ways in which urban contexts are differentially positioned within a wider multilevel governance context that both enables and constrains the development of new decarbonized networks and how local capacity to act is constituted in the context of global South and in particular contextual specificities of cities of sub-Saharan Africa.

4 Africa's urban low carbon energy transition

Work on energy transitions across the towns and cities of Africa are increasingly present across urban studies (Gebreegziabher et al., 2012), particularly in South Africa (Jaglin, 2013; Swilling, 2013) and including engagement with neighbourhood transitions (Bulkeley et al., 2014; Mdluli and Vogel, 2010) and low carbon concerns (Silver, 2014, 2017). However, these remain relatively limited compared with the focus on either household and national-scale transition dynamics, meaning there are still too few accounts that have interrogated energy transition and especially low carbon transition at an explicitly urban scale. Building on this failure to engage with the urban dimension of energy transition is the assumption that towns and cities in the region have little capacity to shape future low carbon transitions compared with other intermediaries such as national governments. We would challenge such assertions as outlined above, arguing that the energy regime intersects in a range of ways with urban innovation, technology development and public pressure, all of which require more sustained scrutiny.

Developing research and associated debates across urban studies can provide an important way to help UTA address the series of emergent low carbon dynamics and imperatives across sub-Saharan African towns and cities. These understandings of urban futures suggest very different trajectories from the urban (modernization) experiences of the global North (Robinson, 2002, 2006). This is a particularly important point in considering the changing nature of infrastructure space and notions of low carbon transition across these urban contexts (Swilling and Annecke, 2012). Importantly, such work prompts us to examine how we might contribute to an expanded UTA by paying close attention to how these regional energy/urban geographies challenge this approach, to better address the particular and geographically located infrastructure conditions of urban Africa (Swilling, 2011).

We now examine work on the diverse geographies across urban Africa to generate four contributions into how such an expanded UTA of low carbon urban transformation can be deployed. These contributions are based on the distinct forms of urbanization being generated, the place of the 'urban' in understanding low carbon transitions in the region, the actors involved in low carbon innovation and the contested and political nature of energy systems across urban contexts as they experience transition. Bringing these concerns together would help UTA to better address how towns and cities beyond the global North shape and are shaped by the system innovation, new technologies, networked relations and broader political-economic transformations implicated in low carbon transition.

Specificity of African urbanization

We draw attention to the particular forms of urbanization across urban Africa that challenge how energy networks are conceived, the notion of a low carbon transition in these contexts and the imperatives of addressing intersections of poverty, informality and broader urban transformation. Infrastructure has tended to be understood through the dichotomy of formality and informality (Varley, 2013) yet new approaches are shifting our understandings to consider alternative forms of infrastructure provision, operation and usage that are multiple and hybrid, consisting of standard hardware, improvised technologies (Silver, 2014) and people themselves (Simone, 2004a). This work draws attention to what Furlong (2014, p. 139) terms the, "coexistence among socio-technical systems, as opposed to the universality of a single dominant infrastructure network" alongside the everyday practices that (re)make infrastructure space across various socio-material configurations (McFarlane, 2009; Silver, 2014; Simone, 2004a). This growing body of literature explores these particular urbanization dynamics covering multiple scales from the everyday through to the broader urban and regional geographies of sub-Saharan Africa (Myers, 2003; Pieterse, 2008, 2010a, 2010b; Simone, 2004a, 2004b, 2010a; Swilling, 2011). Taken together this work illustrates some of the important urban conditions and dynamics for urban transition analysis to engage with when considering how low carbon concerns become materialized across the urban energyscape.

These include the historically 'splintered urbanism' of towns and cities both during and after the colonial era (Graham and Marvin, 2001; Swilling, 2011) including those of sub-Saharan Africa and that, unlike the global North, reveal the historically produced and ongoing fragmented and divided nature of urban (energy) systems in global South contexts. These conditions have favoured colonial, and later postcolonial elites (Bakker, 2003; Furlong, 2014; Jaglin, 2008; Odendaal, 2011; Silver, 2015), opening up a politics concerning which social interests will benefit from the low carbon transition. These historical forms of uneven urban infrastructure provision shape distinct geographies in which the rise of the 'infrastructural ideal' (Graham and Marvin, 2001) offers only a partial narrative of how colonial and postcolonial logics of control, segregation, exploitation and various forms of (under)development mediate unfolding low carbon transitions. They ask us to take seriously the overlapping spatial legacies of colonial, postcolonial (and, of course, in South Africa, apartheid) urban energy governance and the logics of specific colonial–capitalist relations in how decarbonization will proceed. These histories have produced spatial configurations of racialized, gendered and class inequality in accessing basic urban resource provision (Demissie, 2007; Myers, 2006) that are likely to continue and mutate into variegated, unequal experiences of low carbon transition (Silver, 2017).

Related to these historical dynamics are the associated conditions of informality and multiple, urban systems that operate across these infrastructure spaces in the absence of a universal, state-operated infrastructure (Pieterse, 2008; UN–Habitat, 2014). These energy geographies prompt the need to recast how urban conditions are understood in UTA and how they will shape low carbon trajectories. We can draw on some key concerns here including the (overlapping) roles of both formal and informal energy systems in the operation and circulation of energy resources, high levels of energy poverty, reliance on fuels such as charcoal or firewood, together with ongoing struggles over land, tenure and broader service provision (Figure 9.2). These forms of urbanization challenge and stretch global North-anchored

FIGURE 9.2 In Kampala, Uganda, households use charcoal for a range of energy uses beyond the grid.

Source: the authors.

understandings of what constitutes an urban energy network and thus the transition to low carbon operation. As such we would suggest that these differentiated forms of urbanization generate variegated energy geographies across towns and cities that challenge whether any 'modernization' incorporating 'linear' transition to modern energy services and then low carbon is possible or even desired. This is particularly relevant for the large and ever-growing number of informal urban settlements across much of sub-Saharan Africa which often remain unelectrified. As Sokona et al. (2012, p. 5) argue, "[t]he low levels and lack of access to modern energy services for productive activities has also impacted negatively on development and entrenched poverty in the continent." Further issues related to these informal urban spaces include issues about recognition of land (often a precursor to formal electricity connections), the high concentrations of poverty in these areas (in sustaining flows of electricity), high levels of density making interventions difficult to plan, high levels of unauthorized connections and a series of health and safety issues including widespread risks of fire that can quickly destroy people's lives.

Locating the urban context

Second, the location of the 'urban' in energy transition across these towns and cities is a pressing concern in considering how we develop analytical vocabularies to address low carbon futures (Hodson and Marvin, 2010). This is particularly important in considering the institutional limitations across urban authorities in fostering systemic innovation (Agbemabiese et al., 2012). There is important work concerned with how household-scale transitions are occurring (Kowsari and Zerriffi, 2011) and particularly focused on how poverty mediates energy usage/transition (Kebebe et al., 2010; Visagie, 2008). For instance, Karekezi et al. (2008) show that in Kenya, despite urban dwellers requiring access to electricity for modern energy services, a range of off-grid fuel sources and technologies form a key part of everyday energy usage and experience in cities such as Nairobi. These everyday household experiences of energy services are established through poverty and inequality but also the informal conditions of many urban spaces. Hiemstra-van der Horst and Hovorka (2008) challenge how energy transition approaches understand household energy usage through empirical research in Maun, Botswana. By illustrating differentiated household energy use patterns the authors challenge notions of linear pathways to modern fuel consumption and the range of structural and everyday factors that shape household energy decisions and assumptions about energy transition. Furthermore they go on to suggest, "multiple energy sources are employed in complex ways, each for specific purposes, such that modern fuel uptake largely complements fuelwood rather than leading to its abandonment" (Hiemstra-van der Horst and Hovorka, 2008, p. 3342). This work on household-level transition pathways must be incorporated into research on sub-Saharan Africa's urban low carbon transition and emerging studies advancing a second generation of urban transition studies. There are emerging examples that link the urban scale and associated political economies of energy in ways that show how low carbon-focused work might develop such as Baptista's (2016) examination of everyday practices of prepaid electricity in Maputo.

At the national scale these energy transitions are also comparatively interrogated across a number of different countries revealing the diversity of experiences in multiple geographical contexts (Baker et al., 2014; Khennas, 2012; Krupa and Burch, 2011). They offer a moment of caution in seeking to make generalizations and reinforce assumptions across the many urban worlds of sub-Saharan Africa. These studies also reflect the perceived importance of national governments, often at the expense of considerations focused on an urban energy regime in low carbon transition pathways. This focus on the national scale often extends to the constituent regional dynamics and offers important overviews of shared energy geographies across the continent (Brew-Hammond, 2010). This research seeks to consider the key drivers, actors, dynamics and outcomes of energy transitions across the continent that are of course useful in placing and understanding urban conditions even if they often locate the energy regime beyond the boundaries of towns and cities.

Constituting urban capacity for low carbon transition

Third, we contend it is crucial to expand the understanding in UTA of wider social interests involved in low carbon, technological innovation in urban Africa, and the deployment, maintenance and operation of multiple energy systems. In particular we argue that the agency of 'slum dwellers' and associated social movements and civic organizations in informal settlements will be integral to decarbonization. As Silver (2014) suggests infrastructure space in cities such as Accra have long been associated with incremental and ongoing interventions by urban dwellers seeking to transform conditions of energy poverty and wider socio-environmental inequality. And this incremental upgrading has long been important to broader debates within the academy and policy worlds concerned development and urban planning in the global South (Hasan and Chetan, 1986; Satterthwaite and Mitlin, 2013; Turner, 1972). As such it is likely that low carbon concerns will become integrated with current processes of *in situ* upgrading of informal settlements in ways that have yet to be fully examined or conceptualized.

These urban-, neighbourhood- or even household-scale transitions, often termed niches in the socio-technical literatures (Geels, 2002), illustrate the need to better consider the role of social movements and civil society in experimentation, innovation and technological up-scaling of low carbon technologies and systems. As Ferguson (2006) so usefully elucidates, this 'civil society' cannot simply be grounded within the context of the 'local'. As groups such as Slum Dwellers International show, these urban poor movements are intrinsically linked to transnational networks of solidarity, financing and the co-production of knowledges around delivering essential urban service provision (McFarlane, 2009; Satterthwaite and Mitlin, 2013), and potentially, in the future, low carbon objectives. Further attention should of course be paid to the emergence of what commentators term 'the rising Africa middle class' (Melber, 2013; Ravallion, 2010; Visagie and Posel, 2013). While this is a contested term, the growth of suburban developments (Mabin et al., 2013) and new housing geographies (Grant, 2009; Mercer, 2014) together with new consumption habits and technology usage predicate a series of niches from which innovations are being generated and reshaping the socio-technical landscapes of energy across these towns and cities. Our brief outlining of these actors offers some important intermediaries beyond the focus on the elite within UTA.

Contested urban transitions

Finally, the contestations and politics over infrastructure development, service provision, operation, maintenance and repair all critically shape urban low carbon transitions. Most usefully, Lawhon and Murphy (2012) have offered important tools to interrogate the contested and political nature of urban transition processes. This politicized approach to transition shows how and why UTA should explicitly centre on how power, politics and inequalities across the city come to shape the ways in which transition is planned, operationalized and unfolded. Here Lawhon and Murphy (2012, p. 372) suggest, "how political ecology can improve it through a deeper consideration of the role of knowledge, diversity, power, geography, and non-material circumstances in shaping transition dynamics." This provocation seeks to build on transitions research and debates in urban studies that explicitly politicize infrastructures (McFarlane and Rutherford, 2008). This includes work on politics and contestations over these urban systems in African towns and cities (Gandy, 2006; Loftus, 2006, 2012; Myers, 2003; Silver, 2015, 2017) that have emerged specifically from the urban political ecology literatures and we would suggest have a key role to play in developing UTA to account for the politics of low carbon transitions across the region. Specifically, this work would help to interrogate the contested natures of urban energy regimes, paying attention to the ongoing production (and circulation) of inequalities and injustice through and across energy systems together with sites and processes that produce divergent visions of low carbon futures energy transition. These have perhaps been most visible in urban studies through accounts of the struggles over metering installations and wider service delivery tensions between the urban poor, municipalities and utility providers in South Africa (McDonald, 2009; Ruiters, 2007) but may become more focused on low carbon as it becomes central to energy policy.

5 Conclusion

Given the scope and scale of the challenges and pressures involved in Africa's urban low carbon transition there is both a societal and research need to develop explanations that understand both the limits and opportunities for urban governance in shaping sustainable and socially just energy futures. UTA provides an important way to think about the pressures, institutional contexts and forms of experimentation involved in understanding these emerging low carbon geographies, particularly in relation to the central importance of energy systems through which much of this transformation will occur.

Our review has offered four contributions that need to be addressed within UTA to expand its ability to account for the urban African energyscapes shaping low carbon futures. First, we argued for the need to address the relations and tensions across formal and informal practices of energy production, distribution and consumption as part of the urban futures of Africa. This means recognizing that large parts of energy systems sit outside a conventional understanding of transition and standardized integrated and modernist notions of infrastructure configuration. Consequently further work needs to be undertaken in enlarging the range, methods and ways of analysing the informal and its relations with formal networks through notions of hybridity, everyday interactions with the energy system and the incremental nature of urbanism across urban space (Figure 9.3). Second, we showed the need to recognize the limits of urban government and governance in terms of the capacity and knowledge to understand and reshape urban energy systems around low carbon concerns. This means looking more widely at which organizations are active in intervening in the energy systems and how they may produce differing visions of decarbonization and the ways that these urban governance actors are assembled across multiple scales and through multiple relations. Third, we outlined the need to develop an understanding of what forms of intermediary capacity can be created in different urban contexts, particularly the role of energy users in

FIGURE 9.3 Electrification in Kibera, an informal settlement in Nairobi, Kenya.
Source: the authors.

informal settlement contexts. Critical to this is an understanding of the role of international organizations – NGOs, development agencies, universities – that may be active in constituting visions and capacities for externally imposed low carbon transformation. Here, further work is required to incorporate the everyday experiences of infrastructure and the capacity of social movements, civic organizations and neighbourhood groups to become involved in governing low carbon. Finally, we argued for a better understanding of contested and multiple transition pathways being opened in this complex institutional context. In particular we wish to draw attention to the need to explore the politicized relations between different social, ecological and economic outcomes of particular low carbon pathways. Further work needs to centre the importance of politics in the shaping of networked systems and the ways that such dynamics reinforce and reflect current and future power relations that assemble across low carbon transitions.

We are not suggesting that these contributions provide easy answers. They do however provide a means for locating UTA within a series of particular geographical and political-economic contexts. Low carbon transformation will be shaped through urbanization patterns, municipal capacities, regime dynamics, politics and power relations. Through this expansion of UTA to better account for these geographical processes we would be better equipped as researchers to examine what sort of low carbon interventions can be developed at an urban scale and what sort of social interests are included (or excluded). As the climate crisis and urban responses to decarbonize become increasingly integrated into governing systems and technologies across cities, the need to unpack the trajectories of transition will become critical to broader understandings of urban transformation. In this chapter we have responded to the concerns of the volume to consider the next generation of work on low carbon transitions. This second generation of urban transition studies will have to better account for the spatialities, innovations and contested politics that are reshaping energy geographies across both urban Africa and the global South through new registrars and vocabularies that take on-board and advance new critical understandings of cities and low carbon.

Acknowledgements

The writing of this chapter was funded by SAMSET (Supporting African Municipalities in Sustainable Energy Transitions) and is an EPSRC/DFID/DECC-funded project (Grant No. EP/L002620/1). The chapter is based on a shortened and revised version of Silver, J., and Marvin, S. (2016) 'Powering sub-Saharan Africa's urban revolution: an energy transitions approach', *Urban Studies*, 54(4), 847–861.

References

Agbemabiese, L., Nkomo, J. and Sokona, Y. (2012) Enabling innovations in energy access: an African perspective. *Energy Policy*, 47, 38–47.

Baker, L., Newell, P. and Phillips, J. (2014) The political economy of energy transitions: the case of South Africa. *New Political Economy*, 19(6), 791–818.

Bakker, K. (2003) Archipelagos and networks: urbanization and water privatization in the South. *Geographical Journal*, 169(4), 328–341.

Baptista, I. (2016) Everyday practices of prepaid electricity in Maputo, Mozambique. *International Journal of Urban and Regional Research*, 39(5), 1004–1019.

Brew-Hammond, A. (2010) Energy access in Africa: challenges ahead. *Energy Policy*, 38(5), 2291–2301.

Bridge, G., Bouzarovski, S., Bradshaw, M. and Eyre, N. (2013) Geographies of energy transition: space, place and the low-carbon economy. *Energy Policy*, 53, 331–340.

Bulkeley, H., Castán Broto, V., Hodson, M. and Marvin, M. (eds.) (2010) *Cities and Low Carbon Transitions*. London: Routledge.

Bulkeley, H., Luque-Ayala, A. and Silver, J. (2014) Housing and the (re)configuration of energy provision in Cape Town and Sao Paulo: making space for a progressive urban climate politics? *Political Geography*, 40, 25–34.

Coenen, L., Benneworth, P. and Truffer, B. (2012) Towards a spatial perspective on sustainability transitions. *Research Policy*, 41(6), 968–979.

Coenen, L. and Truffer, B. (2012) Places and spaces of sustainability transitions: geographical contributions to an emerging research and policy field. *European Planning Studies*, 20(3), 367–374.

Demissie, F. (2007) Imperial legacies and postcolonial predicaments: an introduction. *African Identities*, 5(2), 155–165.

Eskom (2012) *Electrification Roadmap for South Africa, Africa and Developing Countries*. Available at: http://www.eskom.co.za/OurCompany/SustainableDevelopment/Documents/E101645EskomSustainabilityBrochurelr2.pdf (accessed 12 August 2015).

Ferguson, J. (2006) Transnational topographies of power: beyond the 'State' and 'civil society' in the study of African politics. Draft mimeo. Department of Anthropology, University of California, Irvine.

Furlong, K. (2014) STS beyond the 'modern infrastructure ideal': extending theory by engaging with infrastructure challenges in the South. *Technology in Society*, 38, 139–147.

Gandy, M. (2006) Planning, anti-planning and the infrastructure crisis facing metropolitan Lagos. *Urban Studies*, 43(2), 371–396.

Gebreegziabher, Z., Mekonnen, A., Kassie, M. and Köhlin, G. (2012) Urban energy transition and technology adoption: The case of Tigrai, northern Ethiopia. *Energy Economics*, 34(2), 410–418.

Geels, F. (2002) Technological transitions as evolutionary reconfiguration processes: a multilevel perspective and a case-study. *Research Policy*, 31(8), 1257–1274.

Geels, F. (2010) The role of cities in technological transitions: analytical clarifications and historical examples. In: Bulkeley, H., Castán Broto, V., Hodson, M. and Marvon, S. (eds.) *Cities and Low Carbon Transitions*. London: Routledge, pp. 37–54.

Geels, F. and Schot, J. (2007) Typology of socio-technical transition pathways. *Research Policy*, 36(3), 399–417.

Ghana Energy Commission (2011) Energy statistics 2000–9. Available at: http://www.energycom.gov.gh/pages/docs/energy_statistics.pdf (webpage no longer accessible, accessed 2 May 2014).

Graham, S. and Marvin, S. (2001) *Splintering Urbanism: Networked Infrastructures, Technological Mobilities and the Urban Condition*. London: Routledge.

Grant, R. (2009) *Globalizing City: The Urban and Economic Transformation of Accra*. Syracuse, NY: Syracuse University Press.

Hasan, A.M. and Chetan, V. (1986) Two approaches to the improvement of low-income urban areas – Madras and Orangi. *Habitat International*, 10(3), 225–234.

Hiemstra-Van der Horst, G. and Hovorka, A.J. (2008) Reassessing the "energy ladder": household energy use in Maun, Botswana. *Energy Policy*, 36(9), 3333–3344.

Hodson, M. and Marvin, S. (2010) Can cities shape socio-technical transitions and how would we know if they were? *Research Policy*, 39(4), 477–485.

Hodson, M. and Marvin, S. (2012) Mediating low-carbon urban transitions? Forms of organization, knowledge and action. *European Planning Studies*, 20(3), 421–439.

ICA (Infrastructure Consortium for Africa) (2009) *Annual Report 2009*. Tunis Belvedere: ICA.

IEA (International Energy Agency) (2014) *Africa Energy Outlook*. Paris: IEA.

Jaglin, S. (2008) Differentiating networked services in Cape Town: echoes of splintering urbanism? *Geoforum*, 39(6), 1897–1906.

Jaglin, S. (2013) Urban energy policies and the governance of multilevel issues in Cape Town. *Urban Studies*, 51(7), 1394–1414.

Karekezi, S., Kimani, J. and Onguru, O. (2008) Energy access among the urban poor in Kenya. *Energy for Sustainable Development*, 12(4), 38–48.

Karekezi, S. and Majoro, L. (2002) Improving modern energy services for Africa's urban poor. *Energy Policy*, 30(11), 1015–1028.

Kebede, E., Kagochi, J. and Jolly, C.M. (2010) Energy consumption and economic development in sub-Sahara Africa. *Energy Economics*, 32(3), 532–537.

Khennas, S. (2012) Understanding the political economy and key drivers of energy access in addressing national energy access priorities and policies: African perspective. *Energy Policy*, 47, 21–26.

Kowsari, R. and Zerriffi, H. (2011) Three dimensional energy profile: a conceptual framework for assessing household energy use. *Energy Policy*, 39(12), 7505–7517.

Krupa, J. and Burch, S. (2011) A new energy future for South Africa: the political ecology of South African renewable energy. *Energy Policy*, 39(10), 6254–6261.

Lawhon, M. and Murphy, J. (2012) Socio-technical regimes and sustainability transitions: insights from political ecology. *Progress in Human Geography*, 26(3), 354–378.

Loftus, A. (2006) Reification and the dictatorship of the water meter. *Antipode*, 38(5), 1023–1045.

Loftus, A. (2012) *Everyday Environmentalism: Creating an Urban Political Ecology*. Minneapolis: University of Minnesota Press.

McDonald, D. (ed.) (2009) *Electric Capitalism: Recolonising Africa on the Power Grid*. London: Routledge.

McFarlane, C. (2009) Translocal assemblages: space, power and social movements. *Geoforum*, 40(4), 561–567.

McFarlane, C. and Rutherford, J. (2008) Political infrastructures: Governing and experiencing the fabric of the city. *International Journal of Urban and Regional Research*, 32(2), 363–374.

MacLeod, G. and Jones, M. (2001) Renewing the geography of regions. *Environment and Planning D: Society and Space*, 19(6), 669–695.

Mabin, A., Butcher, S. and Bloch, R. (2013) Peripheries, suburbanisms and change in sub-Saharan African cities. *Social Dynamics*, 39(2), 167–190.

Madlener, R. and Sunak, Y. (2011) Impacts of urbanization on urban structures and energy demand: what can we learn for urban energy planning and urbanization management? *Sustainable Cities and Society*, 1(1), 45–53.

Mdluli, T.N. and Vogel, C.H. (2010) Challenges to achieving a successful transition to a low carbon economy in South Africa: examples from poor urban communities. *Mitigation and Adaptation Strategies for Global Change*, 15(3), 205–222.

Melber, H. (2013) Africa and the middle class. *Africa Spectrum*, 48(3), 111–120.

Mercer, C. (2014) Middle class construction: domestic architecture, aesthetics and anxieties in Tanzania. *Journal of Modern African Studies*, 52(2), 227–250.

Myers, G. (2003) *Verandahs of Power: Colonialism and Space in Urban Africa*. Syracuse, NY: Syracuse University Press.

Myers, G. (2006) The unauthorized city: late colonial Lusaka and postcolonial geography. *Singapore Journal of Tropical Geography*, 27(3), 289–308.

Odendaal, N. (2011) Splintering urbanism or split agendas? Examining the spatial distribution of technology access in relation to ICT policy in Durban, South Africa. *Urban Studies*, 48(11), 2375–2397.

Parnell, S. and Pieterse, E. (eds.) (2014) *Africa's Urban Revolution*. London: Zed Books.

Parnell, S. and Walawege, R. (2011) Sub-Saharan African urbanisation and global environmental change. *Global Environmental Change*, 21(1), S12–S20.

Pieterse, E. (2008) *City Futures: Confronting the Crisis of Urban Development*. London: Zed Books.

Pieterse, E. (2010a) Cityness and African urban development. *Urban Forum*, 21(3), 205–219.

Pieterse, E. (ed.) (2010b) *Counter-Currents: Experiments in Sustainability in the Cape Town Region*. Cape Town: Jacana Media.

Power, M., Newell, P., Baker, L., Bulkley, H., Kirshner, J. and Smith, A. (2016) The political economy of energy transitions in Mozambique and South Africa: the role of the rising powers. *Energy Research & Social Science*, 17, 10–19.

Ravallion, M. (2010) The developing world's bulging (but vulnerable) middle class. *World Development*, 38(4), 445–454.

Raven, R., Schot, J. and Berkhout, F. (2012a) Space and scale in socio-technical transitions. *Environmental Innovation and Societal Transitions*, 4(1), 63–78.

Raven, R., Schot, J. and Berkhout, F. (2012b) Breaking out of the national: foundations for a multi-scalar perspective of socio-technical transitions. Working Paper 12.03. Eindhoven: Eindhoven Centre for Innovation Studies, School of Innovation Sciences, Eindhoven University of Technology.

Robinson, J. (2002) Global and world cities: a view from off the map. *International Journal of Urban and Regional Research*, 26(3), 531–554.

Robinson, J. (2006) *Ordinary Cities: Between Modernity and Development* (Vol. 4). London: Psychology Press.

Rohracher, H. and Spath, P. (2014) The interplay of urban energy policy and socio-technical transitions: the eco-cities of Graz and Freiburg in retrospect. *Urban Studies*, 51(7), 1415–1431.

Ruiters, G. (2007) Contradictions in municipal services in contemporary South Africa: disciplinary commodification and self-disconnections. *Critical Social Policy*, 27(4), 487–508.

Rutherford, J. and Coutard, O. (2014) Urban energy transitions: places, processes and politics of socio-technical change. *Urban Studies*, 51(7), 1353–1377.

Satterthwaite, D. and Mitlin, D. (eds.) (2013) *Empowering Squatter Citizen: Local Government, Civil Society and Urban Poverty Reduction*. London: Routledge.

Silver, J. (2014) Incremental infrastructures: material improvisation and social collaboration across post-colonial Accra. *Urban Geography*, 35(6), 788–804.

Silver, J. (2015) The potentials of carbon markets for infrastructure investment in sub-Saharan urban Africa. *Current Opinion in Environmental Sustainability*, 13, 25–31.

Silver, J. (2016) Disrupted infrastructures: an urban political ecology of interrupted electricity in Accra. *International Journal of Urban and Regional Research*, 39(5), 984–1003.

Silver, J. (2017) The climate crisis, carbon capital and urbanisation: An urban political ecology of low-carbon restructuring in Mbale. *Environment and Planning A*, 49(7), 1477–1499.

Simone, A.M. (2004a) *For the City Yet to Come: Changing African Life in Four Cities*. Durham, NC: Duke University Press.

Simone, A.M. (2004b) People as infrastructure: Intersecting fragments in Johannesburg. *Public Culture*, 16(3), 407–429.

Simone, A.M. (2010a) *City Life from Dakar to Jakarta*. New York: Routledge.

Simone, A.M. (2010b) Infrastructure, real economies, and social transformation: assembling the components for regional urban development in Africa. In: Pieterse, E. (ed.) *Urbanization Imperatives for Africa: Transcending Policy Inertia*. Cape Town: African Centre for Cities, pp. 24–36.

Smith, A., Stirling, A. and Berkhout, F. (2005) The governance of sustainable socio-technical transitions. *Research Policy*, 34(10), 1491–1510.

Sokona, Y., Mulugetta, Y. and Gujba, H. (2012) Widening energy access in Africa: towards energy transition. *Energy Policy*, 47, 3–10.

Spath, P. and Rohracher, H. (2010) 'Energy regions': the transformative power of regional discourses on socio-technical futures. *Research Policy*, 39(4), 449–458.

Spath, P. and Rohracher, H. (2012) Local demonstrations for global transitions – dynamics across governance levels fostering socio-technical regime change towards sustainability. *European Planning Studies*, 20(3), 461–479.

Swilling, M. (2011) Reconceptualising urbanism, ecology and networked infrastructures. *Social Dynamics*, 37(1), 78–95.

Swilling, M. (2013) Economic crisis, long waves and the sustainability transition: an African perspective. *Environmental Innovation and Societal Transitions*, 6, 96–115.

Swilling, M. and Annecke, E. (2012) *Just Transitions: Explorations of Sustainability in an Unfair World*. Cape Town: UCT Press.

Toulmin, C. (2009) *Climate Change in Africa*. London: Zed Books.

Truffer, B. and Coenen, L. (2012) Environmental innovation and sustainability transitions in regional studies. *Regional Studies*, 46(1), 1–21.

Turner, J. (1972) Housing as a verb. In: Turner, J. and Fichter, R. (eds.) *Freedom to Build*. New York: Macmillan, pp. 148–175.

UN–DESA (United Nations Department of Economic and Social Affairs) (2009) Urban and rural areas 2009. Available at: http://www.un.org/en/development/desa/population/publications/pdf/urbanization/urbanization-wallchart2009.pdf (accessed 12 June 2014).

UN–Energy (2005) The energy challenge for achieving the Millennium Development Goals. United Nations Development Programme website, 1 July. Available at: http://www.undp.org/content/undp/en/home/librarypage/environment-energy/sustainable_energy/the_energy_challengeforachievingthemillenniumdevelopmentgoals.html (accessed 12 March 2011).

UN–Habitat (2014) *State of African Cities*. Nairobi: UN–Habitat.

Varley, A. (2013) Postcolonialising informality? *Environment and Planning D: Society and Space*, 31(1), 4–22.

Verbong, G. and Geels, F. (2010) Exploring sustainability transitions in the electricity sector with socio-technical pathways. *Technological Forecasting and Social Change*, 77(8), 1214–1221.

Visagie, E. (2008) The supply of clean energy services to the urban and peri-urban poor in South Africa. *Energy for Sustainable Development*, 12(4), 14–21.

Visagie, J. and Posel, D. (2013) A reconsideration of what and who is middle class in South Africa. *Development Southern Africa*, 30(2), 149–167.

Ward, K. and Jonas, A.E. (2004) Competitive city-regionalism as a politics of space: a critical reinterpretation of the new regionalism. *Environment and Planning A*, 36(12), 2119–2139.

10

LOCALIZING ENVIRONMENTAL GOVERNANCE IN INDIA

Mapping urban institutional structures

Neha Sami

1 Introduction

Urban regions across the world are emerging as critical actors in dealing with climate change (IPCC, 2014; Stone, 2012; IBRD/World Bank, 2010; Betsill and Bulkeley, 2006). This has emerged partly as a result of growing frustration with international negotiation processes and perceived inaction at the level of national governments, as well as due to the realization that the impacts of climate change are going to be increasingly localized and possibly amplified in urban regions (Stone, 2012). As India prepares for a large-scale urban transition and the urban population projected to increase from 377 million in 2011 (Census of India, 2011) to approximately 600 million by 2031, current and future urban populations are increasingly vulnerable to climate change. However, cities and urban regions are also sites of potential opportunity for innovative solutions in dealing with these risks. Focusing on two Indian cities, Bangalore and Chennai, this chapter looks at how these cities and their governments are dealing with the challenge of urban environmental governance more broadly, and climate governance specifically.

Environmental governance in India has historically been underpinned by concerns around reconciling developmental priorities and economic growth with sustainable development pathways (Nair, 2015). Climate change has been largely regarded as a diplomatic rather than developmental or environmental challenge, focusing on preparing for international climate negotiations and climate policy being almost synonymous with India's foreign policy on climate change. (Dubash and Joseph, 2015). "Developing and buttressing legal and conceptual devices to ensure allocations that are in India's interests – understood as ensuring that mitigation efforts did not limit India's options for energy policy and hence for growth – has become the first priority" (Dubash and Joseph, 2015, p. 10). However, over the last decade, there has been a gradual change in perceptions around climate change within national and state governments in India, as well as among various non-state actors including the private sector and civil society groups (ibid.). Since 2007 especially, responding to international pressure for action, there has been a more focused domestic policy movement around climate mitigation and adaptation, including the creation of institutional structures and frameworks at national and state levels to plan and implement climate change action plans.

While growing interest in climate policy and its resulting initiatives are a welcome change, the larger challenge in the Indian context is the creation of an institutional architecture to enable the adoption and implementation of the adaptation and mitigation strategies that are being developed at the national scale. Although a process of creating this framework and the institutionalization of climate policy began during 2007–9, the focus of these initiatives was at the level of the national and state governments. Urban governments have been largely absent in the Indian environmental governance framework, and climate change is no exception. As cities emerge as both locations of climate risk as well as drivers of sustainability (Betsill and Bulkeley, 2006), the absence of Indian urban governments is a serious concern.

Focusing on Bangalore, and Chennai, this research is particularly interested in understanding how a variety of urban stakeholders (state and non-state) participate in and influence climate governance, across scales and sectors in India. Using institutional mapping in both cities – and drawing on primary and secondary research, including interviews with a range of stakeholders and a review of planning documents, government reports and grey literature – this chapter argues that while there are a range of efforts across multiple scales and sectors under way to tackle climate and other environmental challenges, these are often piecemeal and not coordinated. Moreover, there is little engagement from state and local (city) governments in climate governance issues, and a lot of these efforts are an outcome of private or non-state actors and international agencies supporting environmental policy and processes in India.[1] There is no systematic approach at the urban or regional scale to specifically address questions of adaptation or mitigation – most environmental policy is broadly framed and addresses questions of sustainable development rather than focusing specifically on climate change. In the absence of an adequate institutional governance framework, this raises questions about the long-term sustainability of climate action in the Indian context.

The rest of this chapter is organized as follows: the next section builds on earlier and ongoing work on environmental and climate governance, both in India and elsewhere, and provides a broader context for climate policy and governance in India. That is followed by a mapping of the current institutional structure for climate governance, and a discussion of the climate governance framework at the city scale, looking at state and non-state actors and their role in implementing policy at the local level. The conclusion discusses the implications of India's climate governance framework at the urban scale, looking especially at what this means for the country's rapidly growing cities and urban regions.

2 Climate policy and governance in the Indian context

The cornerstone of India's climate policy has been the idea of co-benefits (seen as complementarities across development and climate policy), emphasizing the integration of climate policy into the broader domestic development agenda. India's position on environmental issues has and continues to be largely under-pinned by concerns around reconciling developmental priorities and economic growth with sustainable development pathways (Dubash and Joseph, 2015, p. 9; Nair, 2015). One outcome of this emphasis on co-benefits is the assumption (especially at the subnational level) that adaptation strategies are more desirable than mitigation or low carbon approaches, since they would be more easily amenable to serving the dual goals of development as well as sustainability, and would perhaps be supported more by the Ministry of Environment, Forests and Climate Change (MoEFCC, formerly Ministry of Environment and Forests) (Dubash and Jogesh, 2014). This is true of most environmental planning in India, which emphasizes good sustainable development practices that serve both economic growth as well as sustainability goals.

This means that mitigation solutions and other climate action strategies need to be woven into the larger economic development and growth narrative for the country rather than focusing on specific carbon reduction outcomes or other targeted goals (Parikh et al., 2014). As the report of the Expert Group on Low Carbon Strategies concludes "in such a 'development first' framing, mitigation of GHG [greenhouse gas] emissions is seen as a co-benefit of a sustainable development policy, rather than as the principal objective. Consequently, rather than develop policies specifically for mitigating GHG emissions, the approach prioritizes those development strategies that yield greater decarbonisation – the development imperatives being equal" (Parikh et al., 2014, p. 101). The various national and state Climate Action Plans also build on this notion of co-benefits and main-streaming of climate policy. This focus on co-benefits while favouring adaptation over mitigation is also evident in the state climate action plans that both Karnataka and Tamil Nadu have developed (Sami, forthcoming; Dubash and Jogesh, 2014).

This necessitates the creation of an increasingly complex governance framework around climate change involving integration across multiple existing departments and government agencies (Dubash and Joseph, 2015). Such a task needs a much more systematic approach to institution building than the current *ad hoc* method offers, with roles and responsibilities being integrated across agencies. It also calls for a deeper understanding of the current challenges of developing such a framework at the national level and subnational (state or regional and city) levels. A further challenge is the already messy and complicated urban and regional governance structure in India, characterized by its fragmentation across state, city and parastatal agencies (Weinstein et al., 2013).

Much of the academic and policy writing around environmental governance in the Indian context (and, to a certain extent, the global South) has focused on questions of conflict over resource allocation through governance institutions and the delivery of environmental services (Paavola, 2007; Lemos and Agrawal, 2006; Davidson and Frickel, 2004; Adger et al., 2003). While there has been some work that has looked at the challenges of climate governance for both adaptation and mitigation in the Indian context (Dubash, 2012; Williams and Mawdsley, 2006), this has either largely addressed issues at the national scale or the impacts on specific populations (Byravan and Rajan, 2008, 2006). It is only recently that emerging research has begun to focus on climate governance across scales, with an emphasis on the coming together of the national and state (regional) scales (Dubash and Joseph, 2015; Dubash and Jogesh, 2014).

However, there remains a significant gap in the understanding of the role that Indian cities and urban governments, as well as other non-state actors, can play in this process. Yet, it is not enough to only focus on one particular scale at the sub-national level. Environmental governance issues are not contained by jurisdictional boundaries, often blurring the distinction between different levels, nor do they only affect particular populations. Moreover, governments, their agencies, and officials are far from the only stakeholders – there is a growing involvement of non-state actors including community groups and NGOs, activists, academics and international agencies and networks. A more flexible approach to understanding environmental governance therefore becomes important. In this context, it would perhaps be useful to consider multilevel governance frameworks that emphasize the relations between and across different levels of government as well as a range of state and non-state stakeholders (Corfee-Morlot et al., 2011; Gustavsson et al., 2009; Betsill and Bulkeley, 2006; Bulkeley and Betsill, 2005). This would be especially helpful in the Indian example where governance structures are fragmented, and responsibilities divided between agencies at the state and city scales. It would allow a focus on how competences and authorities are shared between different levels of government. It would also acknowledge that local authorities are not functioning in isolation, but are increasingly participating in governing coalitions that include a range of domestic and international actors (Bulkeley and Betsill, 2005). Finally, it recognizes the role that these networks and actors play in setting the policy agenda and implementation (Gustavsson et al., 2009; and Bulkeley, 2006; Bulkeley and Betsill, 2005). Using this framework to understand environmental governance in India, it becomes apparent that there are several challenges with the participation of local governments in these processes.

While the growing involvement of national-level agencies and state governments in climate policy has been a welcome change, city governments and other urban local bodies have been notably absent from this process. Taking a multilevel governance perspective allows us to examine how the local level alongside networks and governing coalitions between state and non-state actors across different levels of government can influence the implementation and interpretation of environmental sustainability (Bulkeley and Betsill, 2005). Historically, Indian municipal governments have had very little decision-making power and have acted largely as implementation agencies (Weinstein et al., 2013; Pinto, 2000). There has been little change in this, despite several legislative and policy attempts to decentralize urban government (Weinstein et al., 2013; Government of India, 1992). In the case of environmental issues, this is further compounded by the lack of technical capacity within regional and local government agencies to deal with environmental questions as well as the absence of a clear mandate to officials on how to tackle these issues and which local agencies are responsible. Similarly, fragmentation of governance across scales and sectors, alongside a limited ability to take decisions given the concentration of power in regional governments, makes effective local action difficult. Other challenges include the lack of personnel in local or regional governments, a gap in technical capacity to understand and engage with the issues, and an absence of financial resources or the ability/authority to raise them. Few government officials at the local level consider this to be part of their core responsibilities, since climate change was added on to pre-existing functions.

These challenges have also limited various attempts by local stakeholders, international networks and donor agencies to engage with urban governments in India around climate policy. There are a range of domestic and international non-state actors that are actively trying to work on several environmental issues in Indian cities. These actors are actively trying to engage beyond the local level with governments as well as other non-state actors, across national, regional and local jurisdictions, blurring urban, peri-urban and rural boundaries. However, without the participation of urban local bodies, these efforts do not go very far, and effective implementation remains a struggle.

Challenges also remain in terms of different ministries and government agencies at the national level setting up parallel processes and systems, making coordination across different policies and departments very complex. In addition, building on the need to align developmental and environmental priorities, some policies and institutions at national and regional scales have been designed to cut across scales and sectors, while others remained entrenched in specific sectors or departments. However, this, combined with fragmented local governance structures, has posed a particular problem for effective multiscalar governance. Finally, the setting up of climate institutions has been *ad hoc*, and institutions have rarely been stable or long lasting (Dubash and Joseph, 2015). As a result, implementation of climate policy has been piecemeal and incomplete, driven largely by, and dependent on, motivated individuals.

The next section broadly maps the institutional framework across national, state and city scales. This is important to not only understand the institutional architecture for environmental and climate governance at the subnational level in India, but also to identify challenges and opportunities to mobilize for effective climate governance within the current framework.

3 Mapping the institutional framework for environmental governance in India

While there is growing institutionalization of climate policy at the national level, this architecture quickly breaks down at the subnational scale (regional/state and local/urban). This process has not translated into increased capacity or new staff being hired into any of these newly created structures. Technical skills have remained a concern, as has the lack of broader based engagement beyond the government (Sami, forthcoming; Dubash and Joseph, 2015). The national Ministry of the Environment, Forests and Climate Change does not have a strong presence at the subnational level, and regional environment ministries often function independently of the national ministry, leading to a lack of coordination around policy creation and implementation. This is further compounded at the local level, given the lack of power vested in urban governments, a lack of capacity, and the absence of a clear mandate and responsibility to tackle environmental issues.

Local governments have limited ability to raise funds on their own and rely largely on higher levels of government for financial resources, and there has been little funding allocated in the national or state budgets for climate action and related policies. There has been growing national-level interest and action around climate change both internationally and domestically since 2007. However, for climate action to be truly effective, this needs to be reflected at the subnational scale as well as at the state and city levels. This, as we shall see, is more challenging, and one of the key reasons is the lack of a clear institutional architecture around environmental governance issues more broadly and climate governance in particular. The analysis in this section draws on institutional mapping and stakeholder analysis (Aligica, 2006; Smith, 2002) to understand the relative abilities of different climate governance institutions in India to undertake climate action.

National and state (regional) level institutions and climate change

At the national level, MoEFCC is "the nodal agency in the administrative structure of the central government for the planning, promotion, coordination and overseeing the implementation of India's environmental and forestry policies and programmes" (Ministry of Environment and Forests, 2017). The Ministry is divided by sector or area, the responsibility for each of which is assigned to a particular bureaucrat. There are approximately twenty-five Joint Secretaries or Advisers under the environment sector, looking after a range of environmental issues at the national scale. MoEFCC has set up ten regional offices across the country to provide additional environmental management support and advice at the state level (Ministry of Environment and Forests, 2014). These are distinct from the state environmental departments.

At the state level, each state has an environment department, headed by a minister from the state government. These set their own mandate based on the state's priorities. While largely independent, they get some support from the national ministry. Institutionally, state and national environment departments are distinct entities. The institutional structure is distinct from state to state and does not extend to the city level or municipal scale.

Despite the strong presence of the institutional arrangement for environmental governance described above, state-level institutions tasked with climate change are not always aligned to this structure. Since 2010, institutional nodes were established at the state (regional) level, to implement and take forward the State Action Plans on Climate Change (or SAPCCs, mirroring the national NAPCC process; Dubash and Jogesh, 2014). These were meant to be complementary to national climate policy institutions. The state nodal institutions taking on board such responsibilities were not necessarily new agencies or organizations, neither were they always assigned to the pre-existing organizational structure of each state's environment department; in several states climate change was added to the list of responsibilities of already existing institutions, while others created new institutions (ibid.). The Karnataka state government, for example, appointed an autonomous institution under the Department of Forest, Ecology and Environment – the Environmental Management and Policy Research Institute (EMPRI) – as the nodal agency, whereas the Tamil Nadu state government created a new Tamil Nadu Climate Change Cell (TNCCC) to formulate and take forward their plan. The creation of these nodal agencies unfortunately has not resulted in the creation of institutional capacity at the state level to tackle climate change, since there has been little investment in building capacity, hiring technically skilled staff or decentralizing decision-making to these agencies.

These nodal agencies were tasked with the preparation of SAPCCs. Given the lack of technical knowledge and financial resources, most state governments turned to international donor agencies and/or consultants to help with plan preparation (Sami, forthcoming). State nodal agencies, and consequently their plans, have had a weak engagement with climate science. They have also been characterized by a weak framework through which adaptation and/or mitigation strategies can be incorporated into mainstream policy (Dubash and Joseph, 2015). The process of plan formulation itself has been a closed one, mostly conducted by international consultants and with minimal participation from local and regional stakeholders. Despite the engagement with national-level government agencies and several non-state actors (chiefly climate change experts or consultants), state climate institutions do not demonstrate a pattern of multilevel governance. These institutions very much limit themselves to governing specific aspects of environmental issues within their own administrative boundaries.

The remaining sections of the chapter focus on the institutional architecture of climate governance in two case studies at regional and local levels, both of them in southern India: the state of Karnataka and its capital city, Bangalore, and the state of Tamil Nadu and its capital Chennai. Both states share several similar problematics: they are among the most urbanized states in the country, rapidly growing and economically powerful, with a growing service sector economy built around domestic and global networks. In terms of environmental issues, Karnataka's biggest challenge is water, particularly drought and growing depletion of groundwater. Tamil Nadu also has to tackle a range of coastal management issues, with a growing risk of extreme weather events such as hurricanes. Both states also face growing urban environmental challenges, including unsustainable land-use patterns, urban heat island effect, pollution and waste management issues (Sami, forthcoming).

Each state, however, has created its own institutional architecture to respond to these challenges.

Karnataka and Bangalore

In the state of Karnataka, the Department of Forest, Ecology and Environment is responsible for the protection and enhancement of natural resources. It also coordinates and implements environmental acts and rules made by both the national and state governments. The state department has six subregional offices to help with the execution of its responsibilities and monitoring compliance. The emphasis within the state department is largely on forestry, conservation and pollution control. Climate change does not explicitly feature in its mandate. There also seems to be little connection between the state department, its offices, and the regional offices of the MoEFCC (one of which is located in Bangalore). The discussion in the previous section points to an obvious disconnect between the institutional framework at the national and the state levels.

EMPRI officials, the state-level nodal agency to engage with the SAPCC creation process, had a statewide mandate for climate change planning, specifically for the SAPCC process. Their plan development experience was different than most of the other states, since they actually attempted to engage with a range of stakeholders both within government (at the state and city level) and outside. EMPRI's plan drew extensively on academic and scientific knowledge from leading academic institutions in Bangalore. Officials also conducted a range of interviews with other government departments, most of these at the state level, and reached out to a number of local city agencies in Bangalore (Sami, forthcoming; Dubash and Jogesh, 2014). As an institution, EMPRI is largely research and training focused: they have no authority to raise funds or to implement the plan. They also do not fit neatly into MoEFCC's institutional structure, described in the previous section. EMPRI takes direction from the state department, and is dependent predominantly on them for funding. They act in an advisory capacity, and have no enforcement authority. EMPRI expressed frustration with the lack of funding available to implement any of their proposed activities as part of the SAPCC, since they lacked the ability to be able to raise funds themselves. EMPRI officials said that since they already had some funding to test their climate-focused pilot projects, the state was not eligible for more financial support from the national government, and was being encouraged to look for funding from other (private sector or international institutional) sources. As a research institution, it was harder for them to raise funds for project implementation, which would be easier for governments to undertake.

There is very little in Karnataka's SAPCC that focuses explicitly on urban climate governance. According to EMPRI officials interviewed in the context of this research, there was little interest among city officials to engage with climate change–related issues. At EMPRI itself officials come from the forest service and are trained in conservation or forestry for the most part. There was little interest or technical capacity to deal with urban environmental issues broadly, or climate change in particular. EMPRI officials emphasized the lack of capacity as a concern when dealing with urban climate issues (Dubash and Jogesh, 2014). Explaining that their areas of specialization had little urban focus (e.g. agriculture, animal husbandry, horticulture or forestry) they expressed their inability to understand the kinds of climate challenges that urban regions face, particularly because of a lack of training/ skill sets. There are four to five adaptation-related pilot projects being tested, but most of them focus on adaptation strategies in agriculture, horticulture and animal husbandry. Officials at EMPRI were unaware of any specific climate change-related issues being undertaken by city agencies in Bangalore.

In Bangalore, at the city scale, there is no single agency that has a mandate to deal with urban environmental issues, including climate change. Although there is acknowledgement of climate change as an issue at higher levels of the state bureaucracy, this has not yet translated into specific, directed policy action on either adaptation or mitigation. Governance in the city, arguably fragmented, is characterized by a set of functional overlaps among local agencies (Table 10.1). There are several city and parastatal agencies responsible for a range of aspects of environmental governance. For example, there are at least six different city agencies governing Bangalore's roads. There are five key agencies with urban and regional influence and impact whose remit aligns within climate change and low carbon development issues: the Bangalore Development Authority (or the BDA, the city's chief planning agency), the Greater Bangalore City Corporation, or BBMP (Bruhat Bengaluru Mahanagara Palike, the city's municipal corporation), the Bangalore Water Supply and Sewerage Board, the Bangalore Electricity Supply Company, and the Bangalore Metropolitan Transport Corporation. However, across these five agencies there is little acknowledgement of climate change as an urban issue.

There has been considerable interest from a range of international donor organizations to support local climate action in Bangalore, including the Rockefeller Foundation and the Climate and Development Knowledge Network. In addition, there are isolated instances of local environmental groups organizing around specific issues, such as the preservation of green cover or the restoration of Bangalore's lakes, although these are typically reactive and rarely sustain the momentum. These networks of city-based environmental groups play an important role in developing and supporting coordination across state and city levels, through lobbying government agencies to take action or channelling the funds provided by international donor agencies to push state governments towards local implementation. These remain piecemeal efforts. Moreover, according to individuals in donor agencies, working with government agencies at any level is challenging, especially from outside the government. It takes a significant amount of time to build trust within the government. This is often compounded by frequent transfers of officials, and a lack of interest in climate change or environmental issues. In the experience of international donor agencies that have been working with urban local governments in India, it is easier to work with governments with little capacity but interested in tackling these challenges, than to work with a very competent set of officials at a well-resourced government agency but with no interest (Sami, forthcoming).

TABLE 10.1 Bangalore urban governance.

Agency	BBMP	BDA	BMRDA	BWSSB	BESCOM	BMRCL	BMTC	KSRTC	KUIDFC	KSDB
Jurisdiction	Municipal limits	Urban dev. area	Metro area	Metro area	Metro area	Metro area	Metro area	State-wide	State-wide	State-wide
Water, sewerage & drainage	Original		Original	Overlapping						
Electricity					Overlapping					
Transport: road	Original		Original					Overlapping	Original	
Transport: rail						Overlapping				
Road infrastructure	Coordinating	Coordinating	Coordinating						Coordinating	
Development control	Original	Overlapping	Overlapping						Coordinating	
Slum clearance & rehabilitation	Coordinating	Original							Original	Overlapping
Urban finance	Original	Overlapping	Overlapping						Overlapping	
Tax collection	Overlapping									

Legend:
- Original function
- Overlapping function
- Coordinating function
- No involvement

Note: BBMP, Bruhat Bengaluru Mahanagara Palike (or Greater Bangalore City Corporation, the city's municipal corporation); BDA, Bangalore Development Authority; BESCOM, Bangalore Electricity Supply Company; BMRCL, Bangalore Metro Rail Corporation Ltd; BMRDA, Bangalore Metropolitan Regional Development Authority; BMTC Metropolitan Transport Corporation; BWSSB, Bangalore Water Supply and Sewerage Board; KSDB, Karnataka Slum Development Board; KSRTC, Karnataka State Road Transport Corporation; KUIDFC, Karnataka Urban Infrastructure Development & Finance Corporation.

Source: Sami, 2017.

Bangalore has a complex web of state and city agencies, as well as non-profit and citizen groups. If we try to map these institutions in terms of their spatial scale, functionality or power, we get a picture of fragmented governance. In terms of scale, the Department of Forest, Ecology and the Environment and EMPRI both have broad spatial coverage, since their responsibility is for environmental management or climate change planning across the state. However, in terms of functional coverage, both are very narrow. The state department is focused largely on forestry, conservation, pollution control and similar issues, and, for the most part, is concerned with enforcement and compliance. EMPRI is a research institute and its role is limited to knowledge production and dissemination. Given the weak capacity within EMPRI to engage with climate change issues, donor agencies as well as community-based groups prefer to engage directly with the state government rather than with the nodal agency.

At the city scale the challenges are different. Several city agencies with some form of environmental remit are appointed and controlled by state departments. The BDA (the chief planning agency) and the BBMP (the municipal corporation) have jurisdiction within Bangalore city limits, and probably have the widest functional coverage of the city agencies discussed here. However, there is no capacity within the BDA or BBMP to deal with environmental issues. The current draft of the Bangalore Master Plan includes an emphasis on planning for climate change as part of its mandate (Sami, 2017). However, by mid-2017, as this book goes to print, the Master Plan is under litigation and its implementation time frame is uncertain. City utility agencies, also controlled by the state government, have jurisdictional boundaries that are different from each other, making coordination challenging. Their focus is also largely on service delivery. In addition, there are typically multiple agencies involved in governing these services.

Overall, the control and power that the state-level environmental agencies have over local or city-scale climate change issues is limited. Environmental concerns come second to economic development and growth priorities at all levels of government. The state department has the power to make decisions about specific environmental issues and enforce them, but it focuses more on compliance rather than progressive environmental regulations. EMPRI has no authority at all either to make or enforce decisions – the SAPCC that EMPRI prepared therefore is difficult to enforce. Among city-level agencies, the BDA is perhaps the most powerful in terms of decision-making as well as enforcement followed by the BBMP, but they lack engagement or capacity to act. While the utility agencies have a relatively greater degree of power with respect to enforcement, their decision-making ability is weak, since these are largely made at the state level. When it comes to capital and investment capacity, across all state- and city-level institutions, there is little political support for climate change and related action. There is some scepticism around climate change being a urban/regional issue, rather than something that should be resolved at a national and international level.

In summary, in the case of Karnataka and Bangalore, there seems to be little connectivity between national, regional and local institutional frameworks for climate change. The urban is largely missing from the state's plans to address environmental sustainability or climate change. There are few specific institutions charged with climate action apart from the designated nodal agency (EMPRI), which in turn has little power to propose, enforce or implement policy. While there is an attempt being made to align with the eight NAPCC missions launched in 2008 at the national level, the decision regarding which of these missions to adopt is left to the discretion of the state governments. As we shall see, the case is not very different in Tamil Nadu, despite being one of the few Indian state governments to have decentralized urban governance (Sami, forthcoming).

Tamil Nadu and Chennai

In the state of Tamil Nadu the institutional responsibility for environmental governance is divided between two agencies: the Department of Environment and the Tamil Nadu Pollution Control Board. Following the spate of recent weather-related extreme events (from the tsunami in 2005 to the floods of 2015), there have been a series of proposals to tackle natural disasters like floods and hurricanes. However, these are largely reactive, developed in the aftermath of an extreme event, and the disaster management process seems to gradually lose momentum. There is also a regional office of MoEFCC in Chennai. However, as in the case of Karnataka, there seems to be little interaction between this and the state's agencies.

Rather than designating an existing agency, like Karnataka, Tamil Nadu created a new nodal agency for the SAPCC process: TNCCC (the Tamil Nadu Climate Change Cell), based within a local university. Tamil Nadu also enlisted an international consultant – the German Development Agency, GIZ (Deutsche Gesellschaft für Internationale Zusammenarbeit) – to help with the development of the plan. However, as of 2017, there had not been much progress on plan implementation. There was limited information released to the public on the TNCCC's activities or proposed plans. There have been isolated instances of climate action at the regional and city level, led by citizen groups or international agencies such as GIZ, but these have not been coordinated or aligned with the Tamil Nadu SAPCC process.

According to local activists and researchers, climate change is not a priority at the state or city level, despite an acknowledgement of the gravity of the issue from high-level state officials. Consequently, there is little explicit engagement around climate action (either adaptation or mitigation). A recent move towards renewable energy (especially solar) is an exception to this. There is more engagement around specific environmental issues, including a progressive movement around sustainable agricultural practices, and water-related issues (i.e. conservation, flooding and pollution). However, none of these is a coordinated effort, and they end up being piecemeal. Several activists, consultants and academics have expressed frustration with institutions in the state, explaining that although there is a very wide range of institutional actors in Tamil Nadu with a clear and explicit mandate for environmental action, they do not act on it.

At the state level, there are several institutional actors that impact environmental management. These include the State Planning Commission, the Public Works Department, state and district coastal authorities and the Forest and Wildlife Department. Of these, only the coastal authorities and the Forest Department come under the formal authority of the state's Department of Environment. In terms of spatial coverage, almost all of these have a statewide mandate, with the exception of the State Coastal Authority and the district coastal authorities. As with Karnataka, the functional coverage of each agency varies however. The State Planning Commission probably has the broadest functional coverage, since it is responsible for the creation of the state five-year plans, and can mainstream climate policy by factoring it into the plans.

However, it is unclear if or how the State Planning Commission will continue to function, both in view of the dissolution of the National Planning Commission in 2014, and also the political instability in Tamil Nadu from 2016, which by mid-2017 had yet to be resolved. The institution with the next widest functional coverage is the Public Works Department, responsible for construction and maintenance of physical public infrastructure in the state including state buildings, roads and bridges, and the management of water resources. The coastal authorities have medium functional coverage, since they are limited to coastal regulation issues, but can have an impact on a range of issues through policies that they enact. The Forest and Wildlife Department's coverage is similar to that of the State Coastal Authority: restricted spatially and functionally to issues related to and concerned with forests, wildlife and tribal welfare. Almost all of these agencies have limited powers and capacities. They have limited ability to make decisions, and operate largely as enforcing agencies. Most decisions are taken by departments within the state government. However, because several agencies share responsibility for managing a particular resource or function, there are frequent jurisdictional and functional conflicts, which also impact enforcement. Finally, most agencies are weak in terms of both political and financial capacity, with little government funding available to them to be able to implement plans. Finally, with the exception of a few individual officials, there is little buy-in on topics of climate change.

The Chennai story is similar to Bangalore: there is no single agency that has the mandate for urban environmental governance, but rather a range of city and para-statal agencies are responsible for different aspects of environmental management. The Chennai Metropolitan Development Authority (CMDA) and the Chennai Municipal Corporation (CMC) are both responsible for urban planning in the city. The CMDA has wider spatial coverage than the CMC since it is responsible for the greater Chennai urban region. In terms of functional coverage, the CMDA is the chief master-planning authority although the CMC is responsible for land-use planning and solid waste management. There are a few sustainability and low carbon initiatives that have been factored into planning regulations (such as rain-water harvesting or switching to solar water heaters), but these are not part of a larger coordinated effort around climate action.

There is a considerable presence of international donors in Chennai. The city is part of the Rockefeller Foundation's 100 Resilient Cities Programme. The programme has struggled to get a foothold in Chennai for a number of reasons, including a lack of engagement at higher levels of government and a perception by local stakeholders that it was 'corporate funded and led'. GIZ is also closely involved with environmental planning in Chennai, as well as Tamil Nadu, but largely in an advisory capacity (they helped to prepare the SAPCC). Both the 100 Resilient Cities Programme and GIZ are weak on scale and power, but have some political and financial capital. While this has allowed them to engage with environmental issues in the city, they haven't managed to have much impact because of a lack of power to enforce or implement decisions. Both these international agencies have engaged largely with city-level agencies rather than the state government. As a result, they have not been able to get much traction, since the decision-making authority lies with the state and not the city.

There are also several citizen movements in Chennai and surrounding areas. However, these have focused on particular moments or instances (such as the 2015 floods), and have not gained traction beyond a particular neighbourhood or issue. These movements are very limited in terms of scale (functional and spatial) as well as power (although some resident welfare groups are beginning to try and enforce policy within their neighbourhoods). They have moderate political capital across the city, which enables them to mobilize citizens to get involved.

In the case of Tamil Nadu and Chennai, just as in Karnataka and Bangalore, there is a complex web of institutions and actors that have the potential to act but are unable to leverage this potential. One of the chief reasons for this is the influence that electoral party politics has on planning and development in the state in general. When local and state governments are controlled by the same political party, policy-making and implementation is smooth (Sami, forthcoming; Tanner et al., 2009). On the contrary, when there are different political parties in charge, policy-making and implementation breaks down. Interview respondents complained about a 'governance deficit' within the various government institutions at the state and city scale. They lamented the lack of communication and coordination across these agencies, which, they suggested, was one of the reasons why officials were unable to act to enforce their environmental management mandates.

5 Conclusion

By mapping the various institutions involved at the national, state and local levels, this chapter has attempted to delineate the environmental and climate governance architecture in India. There are a range of institutions at multiple scales that have the mandate to act on environmental sustainability, with a growing awareness around climate change in particular. This has manifested itself, at the national and state level, in the creation of a number of new institutions that are responsible for climate action. However, in reality, few of these institutions are stable, and their functioning is *ad hoc* and driven primarily by motivated individuals yet with limited institutional capacity or continuity. In addition, if Indian climate policy is indeed focused on a co-benefits approach, mainstreaming of climate action is essential, and needs a much more knowledge-based integrated approach than is currently the case.

The institutional architecture created at the national level to tackle environmental and climate change issues takes on a different structural form at the state level. While most states have a designated department responsible for environmental issues, these are typically within the state legislature with mandates differing from state to state (as in Tamil Nadu and Karnataka), and have little interaction with the national ministry. National-level institutions rarely have state- or city-level counterparts. This makes continuity in policy difficult, particularly because state agencies have different priorities, as in the case of the NAPCC and the SAPCCs, for example, and restrict themselves to acting within their administrative boundaries. In addition, there is limited engagement with other stakeholders outside of government, such as academic institutions, private sector entities or community groups. International donor agencies, such as the Rockefeller Foundation, and those who work with them at the local level, also experience significant challenges in getting a foothold within regional and local government to be able to influence the implementation of low carbon and other climate or environmental initiatives.

Overall, urban and regional governance in India is fragmented and weak, and there is little engagement with climate policy. Local governments lack personnel and technical capacity, and, as illustrated by Chennai and Bangalore, have not sufficiently leveraged the presence of a range of domestic and international non-state actors. The relationship between state and city government remains top–down, with local agencies responsible for implementation and the decision-making occurring at the state level. The largest challenge, however, is to build support for climate action within state and local government officials. There remains a perception within state and city agencies that climate change is beyond their mandate – an issue that needs to be resolved at a higher scale by national governments. However, in the absence of city or state government agencies taking on a coordinating role for climate action across different scales of government and different sectors, this role is being increasingly taken on by non-state actors such as donor agencies or community-based groups. The challenge is that their influence is limited, and often very focused on a particular problem or sector. Also, without the support of government agencies (especially at the state level), it is almost impossible to move to implementation. Consequently, any investments made in plan development or policy-making do not amount to much if they do not have the support of government agencies at the state level.

Urban governments and their agencies have the potential to play an important role in mobilizing climate action at the city-scale. There are isolated examples of Indian cities and individual champions that are beginning to take action on climate-related issues such as Surat, but these remain few and far between (Sami, forthcoming). There are also opportunities to engage with several non-state actors such as local academics, community organizations, and private sector entities to help fill the technical and knowledge deficit within government and to help mobilize effective climate governance. However, given the Indian governance framework and the lack of power within urban governments, and as the examples of Karnataka and Tamil Nadu show it remains a latent potential.

Acknowledgements

This chapter is based on work funded by Canada's International Development Research Centre and the UK's Department for International Development through the Collaborative Adaptation Research Initiative in Africa and Asia. Parts of this chapter have been previously published as an IIHS (Indian Institute for Human Settlements) commentary on 'Climate Change and the Sustainable Development Goals' for the IIHS Urban Policy Dialogues 2016. Working in seven countries in semi-arid regions, this project seeks to understand the factors that have prevented climate change adaptation from being more widespread and successful, and the processes – particularly in governance – that can facilitate a shift from *ad hoc* to large-scale adaptation. The study sites in India are in Maharashtra, Karnataka and Tamil Nadu. This chapter is focused on two of these three states – Karnataka and Tamil Nadu – and their main urban regions, Bangalore and Chennai. The author would like to acknowledge valuable research assistance by Ritwika Basu, Amogh Arakali and Ankith Kumar. I would also like to thank the book editors, especially Andrés Luque-Ayala, for their very helpful comments.

Note

1 A companion paper to this chapter focuses explicitly on the state and city planning process for climate change in Bangalore and Chennai (Sami, forthcoming).

References

Adger, W.N., Huq, S., Brown, K., Conway, D. and Hulme, M. (2003) Adaptation to climate change in the developing world. *Progress in Development Studies*, 3(3), 179–195.

Aligica, P.D. (2006) Institutional and stakeholder mapping: frameworks for policy analysis and institutional change. *Public Organization Review*, 6(1), 79–90.

Betsill, M. and Bulkeley, H. (2006) Cities and the multilevel governance of global climate change. *Global Governance: A Review of Multilateralism and International Organizations*, 12(2), 141–159.

Bulkeley, H. and Betsill, M. (2005) Rethinking sustainable cities: multilevel governance and the 'urban' politics of climate change. *Environmental Politics*, 14(1), 42–63.

Byravan, S. and Rajan, S.C. (2006) Providing new homes for climate change exiles. *Climate Policy*, 6, 247–252. Available at: https://ssrn.com/abstract=950329 (accessed 14 July 2017).

Byravan, S. and Rajan, S.C. (2008) The social impacts of climate change in South Asia. Available at: http://dx.doi.org/10.2139/ssrn.1129346 (accessed 14 July 2017).

Census of India (2011). Provisional population totals, Census of India, 2011; urban agglomerations/cities having population 1 lakh and above. New Delhi: Census of India. Available at: http://censusindia.gov.in/2011-prov-results/paper2/data_files/India2/Table_3_PR_UA_Citiees_1Lakh_and_Above.pdf (accessed 29 July 2013).

Corfee-Morlot, J., Cochran, I., Hallegatte, S. and Teasdale, P.J. (2011) Multilevel risk governance and urban adaptation policy. *Climatic Change*, 104(1), 169–197.

Davidson, D.J. and Frickel, S. (2004) Understanding environmental governance: a critical review. *Organization & Environment*, 17(4), 471–492.

Dubash, N.K. (ed.) (2012) *A Handbook of Climate Change in India: Development, Politics, and Governance*. New York: Earthscan.

Dubash, N.K. and Jogesh, A. (2014) From margins to mainstream? State climate change planning in India. *Economic & Political Weekly*, 49(48), 86–95.

Dubash, N.K. and Joseph, N.B. (2015) The institutionalisation of climate policy in India: designing a development-focused, co-benefits based approach. Climate Initiative working paper. New Delhi: Centre for Policy Research.

Government of India (1992) *The 74th Constitutional Amendment Act*. New Delhi: Government of India.

Gustavsson, E., Elander, I. and Lundmark, M. (2009) Multilevel governance, networking cities, and the geography of climate-change mitigation: two Swedish examples. *Environment and Planning C: Government and Policy*, 27(1), 59–74.

IBRD/World Bank (International Bank for Reconstruction and Development/World Bank) (2010) *Cities and Climate Change: An Urgent Agenda*. Washington, DC: IBRD/World Bank.

IPCC (Intergovernmental Panel on Climate Change) (2014) *Climate Change 2014: Impacts, Adaptation, and Vulnerability. Part A: Global and Sectoral Aspects. Contribution of Working Group II to the Fifth Assessment Report of the Intergovernmental Panel on Climate Change*. Cambridge and New York: Cambridge University Press.

Lemos, M.C. and Agrawal, A. (2006) Environmental governance. *Annual Review of Environment and Resources*, 31, 297–325.

Ministry of Environment and Forests (2014) *Strengthening and Expansion of Regional Offices under the Central Sector Plan Scheme of Strengthening of Forestry Division – Strengthening of Regional Offices 2013–2014*. New Delhi: Government of India.

Ministry of Environment and Forests (2017) About the Ministry. New Delhi: Ministry of Environment and Forests, Government of India. Available at: http://envfor.nic.in/about-ministry/about-ministry (accessed 4 April 2017).

Nair, J. (2015) Indian urbanism and the terrain of the law. *Economic & Political Weekly*, 50 (36), 54–63.

Paavola, J. (2007) Institutions and environmental governance: a reconceptualization. *Ecological Economics*, 63(1), 93–103.

Parikh, K., Desai, N., Mathur, A., Godrej, J., Banerjee, C., Mathur, R., Bharadwaj, A., Patwardhan, A., Mukerjee, A., Kishwan, J., Tanti, T., Goenka, P., Mathur, J., Choudhury, R., Chandresekhran, I., Pande, V., Sankar, U., Sharma, S.C., Krishnan, S.S. and Chawla, A. (2014) *The Final Report of the Expert Group on Low Carbon Strategies for Inclusive Growth*. New Delhi: Planning Commission, Government of India.

Pinto, M. (2000) *Metropolitan City Governance in India*. Thousand Oaks, CA: Sage.

Sami, N. (2017) Learning by doing: urban planning in Bangalore. In: Vidyarthi, S., Mathur, S. and Agrawal, S. (eds.) *Understanding India's New Approach to Spatial Planning and Development: A Salient Shift?* New Delhi: Oxford University Press, pp. 90–113.

Sami, N. (Forthcoming) Local action on climate change: opportunities and constraints. In: Moloney, S., Funfgeld, H. and Granberg, M. (eds.) *Local Climate Change Action: pathways and progress towards transformation*. London: Routledge.

Smith, C.L. (2002) Institutional mapping of Oregon coastal watershed management options. *Ocean & Coastal Management*, 45(6), 357–375.

Stone, B. (2012) *The City and the Coming Climate: Climate Change in the Places We Live*. New York: Cambridge University Press.

Tanner, T., Mitchell, T., Polack, E. and Guenther, B. (2009) Urban governance for adaptation: assessing climate change resilience in ten Asian cities. *IDS Working Papers*, 315, 1–47.

Weinstein, L., Sami, N. and Shatkin, G. (2013) Contested developments: enduring legacies and emergent political actors in contemporary urban India. In: Shatkin, G. (ed.) *Contesting the Indian City: Global Visions and the Politics of the Local*. Chichester: Wiley-Blackwell, pp. 39–64.

Williams, G. and Mawdsley, E. (2006) Postcolonial environmental justice: government and governance in India. *Geoforum*, 37(5), 660–670.

PART III

Communities and subjectivities

11

GOVERNING CARBON CONDUCT AND SUBJECTS

Insights from Australian cities

Robyn Dowling, Pauline McGuirk and Harriet Bulkeley

1 Introduction

Governance to foster low carbon urban transitions occurs at multiple scales, through myriad means, and diverse institutions. Examples of each abound in the urban carbon governance literature, identifying, for instance, the importance of multilevel governance (Betsill and Bulkeley, 2006), business, government and third-sector actors, and across measures as diverse as formal regulation, financial incentives and provision of low carbon technologies. Across numerous policy domains, the increasing government focus on conducting carbon conduct is acknowledged and explored (Dilley, 2015; Barr and Prillwitz, 2014; Corner and Randall, 2011). There has been considerable policy and scholarly debate on 'behaviour change' governance measures; those designed to engender a low carbon transition by changes in the behaviour of individuals, as householders, workers, travellers and so forth. In Australian cities, for example, a recent audit found that more than two-thirds of a total of 900 local carbon reduction initiatives across Australia's capital cities, for example, had a focus on changing individual action (Mc Guirk et al., 2014). Other key pieces of empirical research have also identified the frequency and intensity of behaviour change mechanisms in urban carbon governance (Moloney et al., 2010; Webb, 2012). Scholarly debate has focused on the efficacy of such governmental programmes, and in particular the assumption of a response being triggered by information (Moloney and Strengers, 2014), as well as critiques of the political implications of such programmes regarding individualization and responsibilization (Whitehead et al., 2011). These discussions are points of departure for our chapter; while they offer important insights they do not exhaust the purchase of such programs in understanding the possibilities for a low carbon transition. Rather than asking whether the subject addressed in behaviour change governance accords with the subject most likely to change, or whether the subject of behaviour change is politically regressive or progressive, we step back to pose a question that must precede such analyses: that is, what subjects are constituted through urban carbon governance and, relatedly, how and by whom are carbon acting subjects constituted. And we do so via an inductive analysis of behaviour change-oriented carbon governance interventions enacted in Australian cities.

Such a focus serves three purposes. First, it acknowledges that governmental programs both act upon and transform subjects (Paterson and Stripple, 2010). In other words, the subject does not simply pre-exist the governmental program, but is formed through it. From this perspective, the question of what is the behaviour-changing subject comes to the fore. Second, investigating subject formation in carbon governance allows us to enrich developing notions of an ecology of urban carbon governance involving a diversity of mechanisms and practices, sites and scales of governing initiatives (McGuirk et al., 2014). Third, it allows us to echo the arguments of Paterson and Mueller elsewhere in this volume (Chapter 12); in particular that addressing socially constituted desires can further low carbon urban transitions. We wish to explicitly explore and connect carbon conduct and subjects to these diverse practices. In essence, the questions for us are threefold: (i) how is the subject addressed in behaviour change governance initiatives?; (ii) how is behaviour rendered governable through these subjectivizing practices?; and (iii) what assumptions of the subject, of the social realm, and of materiality, are enrolled and enacted?

The first section of the chapter sketches the existing scholarship on carbon governance as it relates to the 'conduct of carbon conduct', and signals the ways in which this scholarship is moved laterally by our thinking. In the second section, we draw inductively from our empirical audit of urban carbon governance in Australian cities to identify three forms of subjects and related forms of conduct being worked on through extant behaviour change initiatives – the rational actor, the reflexive, self-disciplining actor, and the materially-embedded actor – and we explicate how and by whom they are being shaped. In the third and concluding section we reflect on the potentials of these subjects and practices for low carbon urban transitions.

2 Governing carbon, conduct and subjects

The foundations of this chapter lie in a Foucauldian governmentality approach that directs attention toward the ways in which governance is a process through which programs and techniques produce desired outcomes through, inter alia, 'the conduct of conduct'. That is, governance entails conducting and constituting subjects, in order to produce desired effects (Dean, 1999). This requires distinct rationalities or mentalities for organizing institutional spaces and the conduct of populations in line with specific aims and objectives (Raco and Imrie, 2000). In their turn, rationalities frame or problematize the objects to be governed, and gather forms of knowledge that make these objects discernible and known in particular ways (Miller and Rose, 1990; Rose, 1996). Effective government, then, relies heavily on ensuring the 'self-government' of relevant actors by constituting subjects whose status, capacities, desires and agencies are imagined and orchestrated to render them governable. Appropriately constituted subjects, by conducting themselves in accordance with rationalities, knowledges and norms aligned with governmental ends, enact governmental objectives (Dean, 1999, pp. 10–11; see also Rutland and Aylett, 2008). As such, subjects are constituted through processes and practices of governance (Bulkeley, 2015). Following Bulkeley et al.'s (2016) excavation of smart grid programs as constituting subjects, we examine the way in which the low(er) carbon subject is constituted and enrolled through the workings of governmental programmes seeking behaviour change.

In addition to an analysis founded in governmentality, we draw upon theoretical frameworks that locate the material world as pivotal in the constitution of subjects. Social practice theory (Shove and Pantzar, 2005; Spurling et al., 2013), assemblage thinking (Müller and Schurr, 2016), and environmental politics (Marres, 2012) are among those that foreground the ways in which objects, material contexts, technologies and the like both shape, and are shaped by, conduct. Social practice theory understands practice as an inherently social and technical/material achievement. It suggests an explicit focus on the role of materiality – the constitutive role of 'things' – in shaping practice (Shove and Pantzar, 2005; Shove et al., 2012), understood as routinized, recurrent forms of behaviour that configure individual and collective conduct (Schatzki, 1996). Assemblage thinking, similarly, attends to the centrality of the material in the various heterogeneous assemblages that constitute specific terrains of action. It suggests both the capacity of particular materialities to produce obdurate outcomes – including particular identifications and subjectivities – and the agentic capacity of the material in reassembling and reshaping such outcomes (Müller, 2015). Finally, literature on material politics (Marres, 2012; Hobson, 2006) has pointed to the capacity of specific objects and material contexts to induce the formation of both pro-environmental subjectivities and behaviours. In drawing attention to the material as a force in the constitution of the subject, these approaches offer complementary conceptualizations on subject formation that can extend and enrich the analytical tools of governmentality approaches.

Drawing on analytics of governmentality and materiality, then, we are especially interested in teasing out the means through which multiple conceptions of the subject are mobilized to render behaviour governable, and by whom it is governed. Thus, ours is not a purely inductive analysis but one guided by existing conceptions of carbon conduct and constituted subjects. Thus informed, we developed a threefold typology summarized in Table 11.1.

The first element of the typology is the notion of the individual as a rational, self-interested decision-maker whose behaviour is determined by her attitudes, values and beliefs. Dolan et al. (2012) term this the 'cognitive model' since changes in an individual's behaviour are understood to be the result of providing the right information that, once processed, generates a shift in the individual's attitudes. Under this approach behaviour change programs are predominantly "enabling and voluntaristic" in nature (Webb, 2012, p. 112), seeking incremental change occurring "through small actions, techno-efficiency measures and 'green' product choices" (Moloney and Strengers, 2014, p. 1). Moloney et al. (2010) note that many pro-environmental behaviour change programs rely on questionable assumptions about how individuals actually react to information, particularly that people will respond rationally to the information presented, and that the 'right' information will necessarily lead to the desired pro-environmental behaviour. In this debate, behaviour change programs are critiqued as being simplistic and misunderstanding of human behaviour and its social and material embeddedness (see Hargreaves, 2011; Shove, 2010). Despite these ongoing critiques, such projects of educating the mind, with their conception of a rational subject, are linchpins in many governance initiatives, including those emanating from urban contexts in Australia, as we illustrate below.

A second subject constituted through behaviour change techniques is the self-governing subject. The disciplining self is much more tightly connected to a Foucauldian conception of governmentality and refers in general to the alignment of self-interest with broader societal interests and the will to improve (see Rose, 1996; summary in Webb, 2012). Thus the 'project' of low carbon transitions becomes incorporated into the ways individuals view themselves and the goals they set. The carbon-calculating individual is thus constituted through governance practices that 'mold and mobilize individual subjects to govern their own emission' (Stripple and Bulkeley, 2015, p. 55). Governing the self is operationalized through numerous mechanisms and practices, including the calculative (carbon diets, competitions), and the financial (carbon pricing, budgets, credits). Critiques of these governance practices also abound, including their radically individualized subjectivity (Paterson and Stripple, 2010; Rice, 2013) connected to a broader trend of individualism and individual responsibilization encouraged by neo-liberal policies and governmental practices (Webb, 2012; Paterson and Stripple, 2010; Fudge and Peters, 2011). Rather than pursue these critiques, we are interested in charting the character and extent of attempts to nurture self-disciplining modes of conduct in the governance of carbon, using the specific example of Australian cities.

A third and final subject relevant to the conduct of carbon conduct is the materially enrolled subject. The theoretical frameworks discussed above foreground the centrality of the material in the constitution of subjects and related 'conduct of conduct'. Material objects like 'bins, bulbs and shower timers' are catalysts and symbols of governmental practices and carbon governance specifically and productive of low(er) carbon subjectivities (Hobson, 2006). Moreover, objects can be instruments of governance, with the deployment of their scripts – "embedded hypotheses about tastes, competences, motives, aspirations" (Akrich, 1992, p. 208) – shaping human conduct in both intentional and unintentional ways (Cloatre and Dingwall, 2013). Thus the materialities of networked infrastructures, washing machines, shower heads, even Nordic walking sticks (Shove and Pantzar, 2005) are deeply implicated in the possibilities, limits and entry points for governing for behaviour change. Spatial layouts and architectures (Jones et al., 2011) are embroiled in the

TABLE 11.1 Typology of behaviour change initiatives, techniques of governance and associated subjects.

Governing intent	Mechanism	Subject
Educating the mind	Demonstrations, information kits, workshops, education leaflets, communication campaigns, environmental expos	The rational subject
Disciplining the self	Audits, commitments/pledges, benchmarking, smart meters, competitions and carbon calculators, financial incentives, building performance ratings, energy monitoring, ecological footprint analysis	The reflexive, self-disciplining subject
Enrolling the 'material'	Solar panels, light bulbs, LPG conversions for cars, hybrid cars, smart meters, energy-efficient appliances	The materially enrolled subject

Note: LPG, liquid petroleum gas.

production of governable low(er) carbon subjects, as are objects like smart meters, zero carbon homes (Walker et al., 2015) and a variety of carbon accounting devices that materialize and thus render energy consumption and/or carbon emission visible and knowable and in turn shape energy demand and emissions (Bulkeley et al., 2016). Importantly, these examples suggest that the materially enrolled subject is constituted in combination with other forms of the subject, a possibility we are alert to as we trace these three forms of subject constitution across Australian urban carbon behaviour change initiatives.

We work through each of these approaches in the remainder of the chapter. Informed by critiques of behaviour change we highlight a persistent and dominant focus on 'educating the mind', with its focus on individuals as rational subjects. We highlight the focus on 'disciplining the self', through a spectrum of self-disciplinary practices (among which neuroliberal-oriented practices (Whitehead et al., 2013) are just one variant), to constitute the reflexive, self-regulating subject. And, guided by the myriad considerations of objects, technologies, spaces, artefacts and networked infrastructures in low carbon urban transitions (Bulkeley et al. 2011), we highlight the importance of engaging materials in orchestrating 'the conduct of (lower carbon) conduct' by the materially enrolled subject. In considering the role of these typologies in carbon conduct and the strategies used to govern it, we acknowledge that these categories are not necessarily discrete nor mutually exclusive. Given the ecology of carbon governance we have identified elsewhere (McGuirk et al., 2015), it is unsurprising to find that the practice of carbon governance enacted through behaviour change is diverse and interconnected; characterized by varied material modes of engagement, involving disciplinary practices that may be rendered neuroliberal, and a variety of educative gestures and informational forms suggesting rational subjects for whom information provision is desirable. As we turn to our empirical analysis, we tease out the character and extent of different forms of behaviour-changing subjects crafted through these diverse practices.

3 Methods

Our analysis of the practices and subjectivities constituted in urban carbon governance is based on an audit of carbon abatement initiatives undertaken by local government areas across all eight of Australia's state and territory capital cities (Sydney, Brisbane, Canberra, Darwin, Adelaide, Melbourne, Hobart and Perth) conducted in 2013. We sampled local governments across these cities to cover small and large, CBD (central business district), inner and mid-city, and outer suburban LGAs (local government authorities). Using local government websites and related material, we identified and documented carbon reduction initiatives related to three domains: buildings, transport and energy infrastructure. Using a framework developed by Bulkeley and Castán Broto (2013), we classified these according to the initiative's focus; who initiated/participated or funded it, the mechanisms the initiative works through, and its target audiences. We identified over 600 unique carbon reduction initiatives, carried out by government, the private sector and non-government agencies individually and in partnership (see review in McGuirk et al., 2014). Of these, 435 sought to govern carbon through governing conduct, and it is these that we focus on here. In what follows we use an extensive and quantitative rather than qualitative or case study approach. This diverges from the predominant means of understanding governmental programs and practices, but does respond to calls to expand the ways in which urban transitions are comprehended (see Bulkeley and Castán Broto, 2013). Moreover, the scale enabled by a quantitative approach, in combination with the breadth of information collected, allows us to draw out the intersections between carbon conduct and other diverse elements of governance. This extensive analysis is then used to discern the character and extent of patterns across governing institutions, techniques and subjects.

A brief overview of these 435 initiatives is in order before moving onto the typological analysis. Local governments (64%) were the main initiators of behaviour change projects, mirroring the general findings of the audit (McGuirk et al., 2014). The low number of federal government-initiated initiatives is noteworthy, as is the small number of community initiatives. Almost half (48%) of the initiatives audited had no partners. When partners were involved, partners were found equally across local governments (29%), corporations (26%) and state governments (5%). But the federal government was noticeably less engaged as a partner on conduct initiatives, at only 6%. Governing carbon conduct in the Australian city focuses on a generalized reduction in energy demand (and by implication – given the dominance of fossil-based energy production in Australia – a reduction in carbon generation), rather than a shift to renewable and therefore low or no-carbon forms of energy. As might perhaps be expected, households and travellers are thus the principal focus of the initiatives constituting this governmental program. This is reflected in the fact that, while initiatives with a behaviour change intent can be found operating across the material domains of energy infrastructure, buildings, and transport, they are particularly dominant in those focused on buildings, whereby householders are encouraged to deploy energy efficiency behaviours to reduce energy demand. Notably though, where initiatives do target the domain of transport, individual mobility (replacing car use with other individual modes like bicycles) rather than mass transit is invariably the focus. On the whole, then, behaviour change can be understood as a governmental strategy that seeks to shape conduct using existing rather than new technologies, using expert/technical knowledge to generate information as a foundation for activities that promote reductions in energy demand and, it follows, carbon emissions.

Consistent with the analytical typology developed above, each behaviour change initiative was coded according to whether it addressed the mind, self or materiality, or some combination of the three. Subjects are multiple rather than singular, and this is also the case for the subjects constituted through urban carbon governance. Thus we explicitly coded for conduct shaped through multiple means, such as educating the mind through materials, or combinations of shaping selves and minds. The typology descriptions outlined in Table 11.1 were used to guide coding. Initiatives that focused on the mind – shaping the rational subject – were coded as those that were primarily educative in nature, encompassing the provision of information (websites, brochures, guides, etc.), workshops, or other functions where the purpose was to educate the subject. Initiatives that disciplined the self – shaping the reflexive, self-disciplining subject – corresponded with those in which calculative and financial techniques were paramount, including self-auditing tools, tools of performance, and practices of accountability (Dean 1999; see also Davoudi and Madanipour, 2013). Initiatives that enrolled the material – shaping the materially enrolled subject – were identified as those working through objects or 'things', such as programs to provide solar panels or other low carbon devices most often introducing, providing and/or focusing on low carbon technologies. We summarize the key ways that conduct was conducted across these initiatives in Table 11.2.

The critical assertion that conduct is principally shaped through information provision – working on 'the mind' – is clearly in evidence. A little more than one-third of initiatives exclusively constituted a rational, information-receiving subject (Table 11.2). Initiatives working on the self and constituting reflexive, self-disciplining subjects were next in importance (21%), followed closely by initiatives enrolling materials to shape subjects (17%). Importantly, while nearly three-quarters of initiatives addressed the subject through just one framing, a quarter combined two approaches (summarized in Table 11.2). However, those that worked through the mind were more likely to do so solely rather than in combination (60 initiatives in combination compared to 151 initiatives through the mind alone). In contrast, when initiatives worked through material, they were much more likely to do so in combination. Indeed, 20% enrolled the rational self with materials, or with self-disciplining processes. It was rare for initiatives to bring together all three foci (3%).

TABLE 11.2 Typology of behaviour change.

Type	Number	%
Mind	151	35
Mind and material	20	5
Self	90	21
Self and material	46	11
Self and mind	40	9
Material	74	17
Self, mind, material	14	3

An important observation to emerge from this summary is that the carbon subject is multiple rather than singular: the information-receiving subject predominates, but coexists with reflexive and materially enrolled subjects. Thus, while critiques regarding self-responsibilization or the limitations of information provision as a basis for shaping carbon conduct remain valid, in focusing on just one form of carbon conduct they miss the complexity of the situation and in particular the relative importance (in terms of frequency) of the conduct being critiqued. In the remainder of the chapter we elaborate on the principal characteristics of each of these forms of subjectivity as modes of governing carbon conduct. To do so, we rely on an analysis of the information presented in Figures 11.1–11.4, which delineate the targets and institutional composition of each. A strict quantitative analysis is not pursued because of our intent to identify and tease out the broad contours of governance through subjectivization rather than explicitly delineate its minutiae. As such, we focus on highlighting the distinguishing features of each form of subjectivity we identify.

4 Educating the mind: the rational subject

As we introduced above, our analysis suggests the most common carbon configuration features a rational subject constituted as in need of education and information in order to develop rational responses that will initiate and guide individuals in changing their patterns of carbon consumption. Initiatives with this focus are principally undertaken by local government acting alone (60%). Where they do emerge through partnerships it is with other local governments and the private sector, rather than with third sector organizations. For such local authorities, educating the mind is the most actionable, being closely aligned with their current expertise and past practice, especially in environmental education and householder engagement. Thus individuals as householders and travellers are the key target audiences. Workshops, information, leaflets, tool kits and expos are targeted at residents as householders (40% of 'mind' initiatives, Figure 11.1); or at travellers (20% of 'mind' initiatives, Figure 11.1). Examples include an outer suburban Sydney local authority's Green Style Your Home initiative that provides a variety of information on energy efficiency, Perth's Canning River Eco-education Centre,

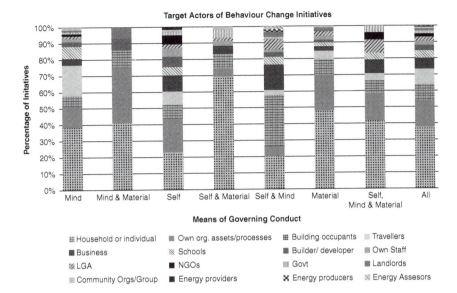

FIGURE 11.1 Target actors of behaviour change initiatives.
Note: LPG, liquid petroleum gas; NGO, non-governmental organization.
Source: the authors.

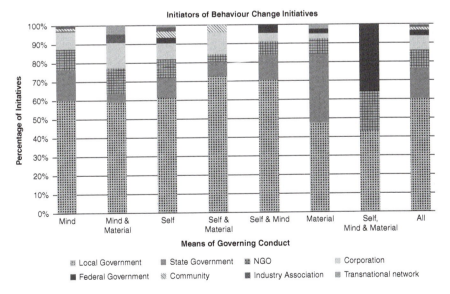

FIGURE 11.2 Initiators of behaviour change initiatives.
Note: NGO, non-governmental organization.
Source: the authors.

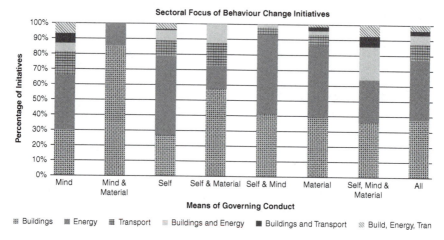

FIGURE 11.3 Sectoral focus of behaviour change initiatives.
Source: the authors.

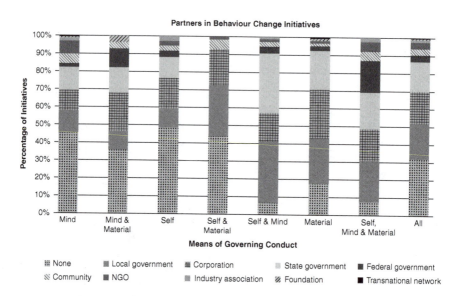

FIGURE 11.4 Partners in behaviour change initiatives.
Note: NGO, non-governmental organization.
Source: the authors.

a facility with a range of activities and demonstrations that aim to raise household awareness of carbon reduction possibilities, or Sydney's Randwick Eco-living Fair (Figure 11.5). 'Travelling subjects' was another pronounced focus here, with a proportionally large number of initiatives focused on supporting an information-deficient travelling subject. These include the provision of information on sustainable transport alternatives, with a concentrated effort on encouraging children to walk or cycle to school. This suggests that transport remains most strongly wedded to a rational notion of behaviour, and a subject- rather than object-focused understanding of travel. The information-receiving subject hence predominated in governing the conduct of carbon conduct in Australian cities, mirroring the characteristics of other parts of the world (Webb 2012; Young and Middlemiss, 2012).

Importantly, there is an additional focus on constituting collective as well as individual rational subjects. Minds were often engaged in collective as well as individual contexts, especially through the provision of information in group situations like workshops. While it is certainly the case that such activities are a cost-effective method of delivering information, it is also the case that they deliver information in a social context. The socially engaged subject, as Hobson and Niemeyer (2011) point out, has merits in carbon reduction strategies, creating shared norms and perhaps enabling the formation of collective or organizational low(er) carbon subjects. Moreover, the contexts for shaping collective subjects work through a distinctive materiality that seeks to educate and constitute subjects via immersive, object-rich and collective material engagements such as demonstration houses and environmental expos (see Whitehead, 2011). Thus, notwithstanding extant critiques of information-provision modes of behaviour change governance, we suggest that the practice of governing carbon conduct through educating the mind is both richly textured and complexly rendered.

5 Disciplining the self: the reflexive self-disciplining subject

Initiatives that work on the self to produce reflexive, self-disciplining subjects comprised 21% of the behaviour change initiatives we identified. They worked largely through technologies of performance (Dean, 1999; see also Davoudi and Madanipour, 2013) that allowed subjects to calculate and thus reflexively regulate their carbon-generating behaviours. These were driven by economically calculative incentives such as financial rewards for low carbon conduct, or through cognitively calculative initiatives in which the impacts of carbon-producing behaviours were made visible and quantifiable, for example via energy audits. Energy conduct was the disproportionate focus here (50% compared to 40% overall, as illustrated in Figure 11.3). Similar to initiatives characterized as educating the mind, this calculative, self-regulating subject was shaped through initiatives with two distinct profiles – those targeting residents and households, and those targeting organizations – constituting both individual and collective/organization low(er) carbon subjects.

Turning first to residents and households, initiatives working on 'the self' were much less likely than mind-focused initiatives to engage individuals as residents or householders. Nonetheless, when residents were engaged as self-disciplining subjects, they were addressed through, for example, eco-challenge events that encouraged residents from specific place-based communities to commit to energy reduction targets as part of a competition, or systems to support public pledges of energy reduction (such as the City of Newcastle's ClimateCam; see Figure 11.6). There are echoes here of shifts in practices of responsibilization from individuals to communities (see Hörnqvist, 2010). Further, these initiatives utilized both moral motivations as the impetus to change, for example via publicly stated value commitments or allegiance to place-based community, and via comparison with others through energy consumption tools. A range of financial mechanisms were also in evidence, such as rebates and subsidies to purchase solar panels, which work through economic modes of calculation and financial motivations. When householders are addressed as reflexive, self-disciplining subjects, it is more likely to be in combination with materials (40 initiatives) rather than disciplinary/calculative practices alone (34 initiatives). Thus through both economically rational, and non-economic affective forms of calculating, reflexive subjects were constituted.

FIGURE 11.5 Randwick Eco-living Fair as an example of educating the mind.
Source: the authors.

FIGURE 11.6 Newcastle ClimateCam as an example of disciplining the self.
Source: the authors.

The second profile emerges around the self-disciplining organization. This arises from the focus of initiatives on organizations and building occupants and businesses (Figure 11.1), in conjunction with partnerships with corporations (Figure 11.4). In contrast to educating the mind, the constitution of the reflexive, self-disciplining subject was more likely to be deployed by organizations (mainly local authorities, but also non-governmental organizations (NGOs) and businesses; Figure 11.2), principally concerned with changing behaviour *within* their own organization as an organizational subject (Figure 11.1). In this vein we see the engagement of techniques like building performance rating systems for organizational buildings, ecological footprinting of organizational assets, or energy monitoring techniques across a set of organizational assets. In the commercial sector, the provision or facilitation of energy audits (in isolation) was common. The key point here however is that an organizational entity was being constituted as the behaviour-changing subject, not an individual, and not, directly at least, workers within those organizations. This finding confirms a point we have recently made elsewhere (McGuirk et al., 2015): that the self-regulation of collective entities – the place-based community or the organization (local government or corporate) – may be reflective of a broadening of engagement beyond the individual in carbon governance, including through behaviour change as a mode of governance.

This focus on the organization as subject is also evident in those initiatives that worked on both minds and selves in combination. This combination made up 9%

of behaviour change initiatives, with almost half of these focusing on the commercial sector (building occupants, businesses, see Figure 11.1). Here we see initiatives like CitySwitch, in which businesses participated in conventional educational techniques like workshops alongside organization-wide energy monitoring and building performance-monitoring practices. This suggests a propensity to construct organizational subjects in distinctive ways – as amenable to a combination of educative and calculative techniques, the latter perhaps not surprising given the more prevalent use of financial tools, audits, etc., in the commercial realm. Moreover, when organizations are enrolled as self-disciplining subjects, it is largely in non-material ways: through self alone or through a combination of self and education.

6 Enrolling the material: the materially-embedded subject

Initiatives that worked through enrolling the material to constitute materially-embedded subjects had a number of distinctive characteristics. First, only half of these material-enrollment initiatives did so in isolation from mechanisms to engage either the self or the mind (74 of the 154 material initiatives worked in isolation). These were typically initiatives that gave away, or exchanged, objects like low-energy light bulbs, or that sought to embed energy-efficient appliances in householders' spaces. That is, technology was enrolled as the vector for shaping behaviour-changing subjects, emblematic of the interplay of scripts and objects in governance and regulation (Cloatre and Dingwall, 2013). These techniques underpin the higher than average focus on households as the target audience of these initiatives (48%, see Figure 11.1). As a set of practices, initiatives that focused on materials were much more likely to be implemented by state governments (more than double the percentage of any other actor in this category), perhaps because of the capacity of state governments to afford higher cost, material-focused and complex projects. Relatedly, this was also the most common mode of behaviour change governance undertaken by corporations, again suggesting the enrolment of technology as a vector to induce behaviour-changing subject formation. Reflecting the distributed agency of a material assemblage, these initiatives were also predominantly constituted through partnerships (80%), forged across local authorities, state governments and corporations (see Figure 11.4). The subject is enrolled in these initiatives via a material intervention: and, crucially, the processes through which objects translate into altered behaviour are assumed rather than directly shaped. There are interesting implications for the debate over individual versus social practices as the more effective entry point for governing carbon conduct (Shove 2010; Whitmarsh et al., 2011) that need further exploring here. The material entry point of these initiatives is contra the critique that, in focusing on individuals, behaviour change initiatives miss out on the central role of the material as elements of practice. However, these initiatives appear to be working through the material without a realized subject, and hence may not be working through practice. Further more detailed qualitative empirical investigation is needed to investigate this potential.

Second, of the other half of material initiatives (80 in total, see Table 11.2) that enrolled materials in combination, operated with mind (9%), self (5%) or mind and self (3%). These combinations vary according to target audience. For initiatives targeting householders, calculative mechanisms working on the self, like the provision of home energy audits, worked in combination with the material objects that enabled subjects to implement the results of these audits (e.g. light bulbs, meters, insulation). When targeting organizations and businesses, materials were provided in conjunction with some form of information provision. Examples here include retrofitting buildings or installation of low carbon technologies (principally solar photovoltaics) and use of the resultant building as a demonstration project, such as the Blue Mountains Visitor Centre (discussed in detail in Dowling et al., 2016). In this respect, the material itself (and in particular the building) becomes both an ends and a means of governance: transformed and then enrolled as a demonstration to entice subsequent behaviour change.

A third characteristic is the relative insignificance in our sample of initiatives that combined materials with both self and mind (3%). It is rare for behaviour change governance to engage all three simultaneously. Australia's national government and NGOs were each at their most active in this complex space (40% of all federal government initiatives addressed all three), with the majority of these being large, multistate initiatives of which behaviour change was just one component; 40% of these focused on the household (Figure 11.1), such as a number of solar city projects that combined financial inducements to install solar, access to purchase solar, and education on energy reduction.

7 Conclusion

Our analysis suggests that just as there is an ecology of carbon governance initiatives in the Australian city (see for example Chapter 7, this volume) there is an ecology of governing carbon conduct. This is operationalized differentially by differently positioned governance actors, sometimes working collaboratively, and enacted across diverse combinations of mechanisms to mobilize subject formation as a site of governance towards low carbon transition. Yet this analysis also suggests and characterizes the multiple forms of behaviour-changing subject that are constituted through governing practice. Different, interconnected subjects are imagined, addressed and thus constituted in relation to particular practices of governance. Low(er) carbon behaviour is governed through educating the mind, assuming a rational subject for which information will instigate a rational (low(e)r carbon) shift in behaviour. Alongside education, and sometimes simultaneously, the behaviour-changing subject is addressed as reflexive and self-disciplining, both responsibilized and enabled to instigate a shift to lower carbon behaviour through calculative and financial techniques – audits, competitions, and other means of measurement – that work across the economic to the moral. And finally, the subject is addressed as responsive to objects like light bulbs, efficient fridges or solar panels, that enable behaviour change by engaging with material aspects of social practice. As a result, there are multiple different kinds of low carbon energy transition being called for and called into being.

Governing for low(er) carbon behaviour change is thus both complex and textured. While our analysis here attempts to characterize the broad landscape evident across behaviour change governing initiatives in Australian cities, deeper understanding of this complex ecology will require further in-depth qualitative analysis. Our extensive analysis nonetheless suggests that more nuanced attention to the constitution of the behaviour-changing subject may reopen some questions around the potential of governing through behaviour change to advance low carbon urban transitions. On the one hand, our analysis confirms well-recognized critiques of governing through behaviour change: a focus on the individual as the behaviour-changing subject, and a related assumption that this individualized subject is rational and calculating. Yet our analysis also foregrounds two dimensions of the subject of behaviour change that point to new paths.

The first is that the subject of behaviour change is not necessarily the individual: it is also the organizational 'self' that inhabits and transforms buildings and is amenable to governance through technologies of performance. Additionally it is the collective 'self' that is also amenable to governance enacted through moral motivations and the mobilization of allegiance to place-based communities. Behaviour-changing mechanisms thus interweave individual, organizational, and collective forms of subject, drawing on rational, calculative techniques alongside those appealing to non-economic, social and affective dimensions, less easily categorized as the rational or the calculative. Yet further, our analysis reveals the imbrication of the material in the constitution of the behaviour-changing subject, pointing to the acknowledgement in many of governing initiatives of the subject as socio-material and embedded in social practice. The Australian urban context demonstrates extant modes of governing behaviour change through subject formations that engage beyond the individual, beyond the rational and, in fact, engage to some degree with the complexities of social practice. Herein, perhaps, lies the potential for mobilizing new productive entry points into governing behaviour change for low carbon transition.

References

Akrich, M. (1992) The de-scription of technical objects. In: Bijker, W. and Law, J. (eds.), *Shaping Technology/Building Society*. Cambridge, MA: MIT Press, pp. 205–224.
Barr, S. and Prillwitz, J. (2014) A smarter choice? Exploring the behaviour change agenda for environmentally sustainable mobility. *Environment and Planning C: Government and Policy*, 32, 1–19.
Betsill, M.M. and Bulkeley, H. (2006) Cities and the multilevel governance of global climate change. *Global Governance: A Review of Multilateralism and International Organizations*, 12(2), 141–159.
Bulkeley, H. (2015) *Accomplishing Climate Governance*. Cambridge: Cambridge University Press.
Bulkeley, H. and Castán Broto, V. (2013) Government by experiment? Global cities and the governing of climate change. *Transactions of the Institute of British Geographers*, 38(3), 361–375.

Bulkeley, H., Castán Broto, V., Hodson, M. and Marvin, S. (eds.) (2011) *Cities and Low Carbon Transitions*. London and New York: Routledge.

Bulkeley, H., Powells, G. and Bell, S. (2016) Smart grids and the constitution of solar electricity conduct. *Environment and Planning A*, 48(1), 7–23.

Cloatre, E. and Dingwall, R. (2013) 'Embedded regulation': the migration of objects, scripts and governance. *Regulation and Governance*, 7(3), 365–386.

Corner, A. and Randall, A. (2011) Selling climate change? The limitations of social marketing as a strategy for climate change public engagement. *Global Environmental Change*, 21(3), 1005–1014.

Davoudi, S. and Madanipour, A. (2013) Localism and neo-liberal governmentality. *Town Planning Review*, 84(5), 551–561.

Dean, M. (1999) *Governmentality: Power and Rule in Modern Society*. London: Sage.

Dilley, L. (2015) Governing our choices: 'proenvironmental behaviour' as a practice of government. *Environment and Planning C: Government and Policy*, 33(2), 272–288.

Dolan, P., Hallsworth, M., Halpern, D., King, D., Metcalfe, R. and Vlaev, I. (2012) Influencing behaviour: the mindspace way. *Journal of Economic Psychology*, 33(1), 264–277.

Dowling, R., McGuirk, P. and Bulkeley, H. (2016) Demonstrating retrofitting: perspectives from Australian local government. In: Hodson, M. and Marvin, S. (eds.) *Retrofitting Cities: Priorities, Governance and Experimentation*. London: Routledge, pp. 212–231.

Fudge, S. and Peters, M. (2011) Behaviour change in the UK climate debate: an assessment of responsibility, agency and political dimensions. *Sustainability*, 3(6), 789–808.

Hargreaves, T. (2011) Practice-ing behaviour change: applying social practice theory to pro-environmental behaviour change. *Journal of Consumer Culture*, 11(1), 79–99.

Hobson, K. (2006) Bins, bulbs, and shower timers: on the 'techno-ethics' of sustainable living. *Ethics, Place & Environment*, 9(3), 317–336.

Hobson, K. and Niemeyer, S. (2011) Public responses to climate change: the role of deliberation in building capacity for adaptive action. *Global Environmental Change*, 21(3), 957–971.

Hörnqvist, M. (2010) *Risk, Power and the State: After Foucault*. London: Routledge.

Jones, R., Pykett, J., and Whitehead, M. (2011) Governing temptation: changing behaviour in an age of libertarian paternalism. *Progress in Human Geography*, 35(4), 483–501.

McGuirk, P., Bulkeley, H. and Dowling, R. (2014) Practices, programs and projects of urban carbon governance: perspectives from the Australian city. *Geoforum*, 52, 137–147.

McGuirk, P., Dowling, R., Brennan, C. and Bulkeley, H. (2015) Urban carbon governance experiments: the role of Australian local governments. *Geographical Research*, 53(1), 39–52.

Marres, N. (2012) *Material Participation: Technology, the Environment and Everyday Publics*. London: Palgrave Macmillan.

Miller, P. and Rose, N. (1990) Governing economic life. *Economy and Society*, 19(1), 1–31.

Moloney, S., Horne, R. and Fien, J. (2010) Transitioning to low carbon communities – from behaviour change to systemic change: lessons from Australia. *Energy Policy*, 38(12), 7614–7623.

Moloney, S. and Strengers, Y. (2014) 'Going green'?: the limitations of behaviour change programmes as a policy response to escalating resource consumption. *Environmental Policy and Governance*, 24(2), 94–107.

Müller, M. (2015) Assemblages and actor-networks: rethinking socio-material power, politics and space. *Geography Compass*, 9(1), 27–41.

Müller, M. and Schurr, C. (2016) Assemblage thinking and actor-network theory: con-junctions, disjunctions, cross-fertilisations. *Transactions of the Institute of British Geographers*, 41(3), 217–229.

Paterson, M. and Stripple, J. (2010) My Space: governing individuals' carbon emissions. *Environment and Planning D: Society and Space*, 28(2), 341–362.

Raco, M. and Imrie, R. (2000) Governmentality and rights and responsibilities in urban policy. *Environment and Planning A*, 32(12), 2187–2204.

Rice, D. (2013) Beyond welfare regimes: from empirical typology to conceptual ideal types. *Social Policy & Administration*, 47(1), 93–110.

Rose, N. (1996) Governing 'advanced' liberal democracies. In: Barry, A., Osborne, P. and Rose, N.S. (eds.) *Foucault and Political Reason: Liberalism, Neoliberalism and Rationalities of Government*. London: UCL Press, pp. 37–64.

Rutland, T. and Aylett, A. (2008) The work of policy: actor networks, governmentality, and local action on climate change in Portland Oregon. *Environment and Planning D*, 26(4), 627–646.

Schatzki, T.R. (1996) *Social Practices: A Wittgensteinian Approach to Human Activity and the Social*. Cambridge: Cambridge University Press.

Shove, E. (2010) Beyond the ABC: climate change policy and theories of social change. *Environment and Planning A*, 42(6), 1273–1285.

Shove, E. and Pantzar, M. (2005) Consumers, producers and practices: understanding the invention and reinvention of Nordic walking. *Journal of Consumer Culture*, 5(1), 43–64.

Shove, E., Pantzar, M. and Watson, M. (2012) *The Dynamics of Social Practice: Everyday Life and How It Changes*. London: Sage.

Spurling, N., McMeekin, A., Shove, E., Southerton, D. and Welch, D. (2013) Interventions in practice: re-framing policy approaches to consumer behaviour. In: *Sustainable Practices Research Group Report*. Sustainable Practices Research Group Report, September. Available at: http://www.sprg.ac.uk/uploads/sprg-report-sept-2013.pdf (accessed 1 August 2017).

Stripple, J. and Bulkeley, H. (2015) Governmentality. In: Bäckstrand, K. and Lövbrand, E. (eds.) *Research Handbook on Climate Governance*. London: Edward Elgar, pp. 49–60.

Walker, G., Karvonen, A. and Guy, S. (2015) Zero carbon homes and zero carbon living: sociomaterial interdependencies in carbon governance. *Transactions of the Institute of British Geographers*, 40(4), 494–506.

Webb, J. (2012) Climate change and society: the chimera of behaviour change technologies. *Sociology*, 46(1), 109–125.

Whitehead, M., 2011. *State, Science and the Skies: Governmentalities of the British Atmosphere*. Oxford: John Wiley & Sons.

Whitehead, M., Jones, R. and Pykett, J. (2011) Governing irrationality, or a more than rational government? Reflections on the rescientisation of decision making in British public policy. *Environment and Planning A*, 43(12), 2819–2837.

Whitehead, M., Jones, R. and Pykett, J. (2013) Neuroliberal climatic governmentalities. In: Stripple, J. and Bulkeley, H. (eds) *Governing the Climate: New Approaches to Rationality, Power and Politics*. New York and Cambridge: Cambridge University Press.

Whitmarsh, L., O'Neill, S. and Lorenzoni, I. (2011) Climate change or social change? Debate within, amongst, and beyond disciplines. *Environment & Planning A*, 43(2), 258–261.

Young, W. and Middlemiss, L. (2012) A rethink of how policy and social science approach changing individuals' actions on greenhouse gas emissions. *Energy Policy*, 41, 742–747.

12

CULTURAL CONFLICTS AND DECARBONIZATION PATHWAYS

Urban intensification politics as a site of contestation in Ottawa

Matthew Paterson and Merissa Mueller

1 Introduction

In many cities, the politics of urban density is becoming increasingly intertwined with strategies for dealing with climate change. While it has long been known among urban climate change advocates, as well as others working for cities that are 'smart', 'liveable', 'sustainable' or some other byword for less car-dependent, that urban density is one of the key determinants of greenhouse gas (GHG) emissions, climate action in most cities has shied away from tackling this question head-on. This aspect of urban climate politics is particularly challenging in North American and Australian cities where the pattern of urban development since the late nineteenth century has produced particularly low density urban forms that are highly car-dependent – 'sprawl' (e.g. Newman and Kenworthy, 1989). Tackling low density 'sprawl' is thus crucial for pursuing low carbon urban development (e.g. Ala-Mantila et al., 2013; Bart, 2010; Glaeser and Kahn, 2010; Hoornweg et al., 2011; VandeWeghe and Kennedy, 2007)

At the same time, since the 1990s, many cities in North America have been attempting to tackle the density question for different reasons. Low density urban development, or 'sprawl', generates a number of other social, economic and environmental costs, in terms of the reduced access to services for those without access to a car, health effects associated with the lack of physical mobility, fiscal costs to municipalities in terms of provision of services (water, sewer, electricity, schools, etc.), and urban air pollution. The recognition of these costs have generated a range of initiatives designed to reorient development back towards inner-urban areas, from changing zoning rules and imposing extra costs on developers for building on the urban fringe, to partnership arrangements for inner-urban construction projects. In many contexts, this process is framed as a question of 'intensification', a frame that has become institutionalized in the municipal planning process. This chapter focuses on one such site, the city of Ottawa – within the Province of Ontario, where the planning basic rules are set out for municipalities like Ottawa – where intensification has become an integral part of the language of planners and local politicians.

These agendas are starting to come together, as we show below, but for the most part we focus here on the politics of intensification as a way to think about the sub-politics of low carbon transitions. That is, generally, the principal justification given for intensification is not explicitly in relation to climate change strategies or low carbon urban transitions (although such references are growing in official rationales), but rather with respect to social cohesion and urban liveability, the fiscal costs of low density development (mostly to do with service provision), enabling transit to work effectively, or smart economic development. We see such interventions as nevertheless part of the politics of low carbon urban transitions: much of this politics goes on in hidden, subtle ways, and does not need to be articulated explicitly in relation to climate change in order for us to consider it as part of urban climate politics. As such, we can learn something about the politics of those transitions by exploring the politics of intensification.

An underlying premise is that while the existing literature on low carbon transitions has tended to focus on innovations, experiments, low carbon projects, etc., we need also to explore ongoing contradictions between existing cultural identities and practices (and the political-economic processes they are closely connected to) one the one hand, and attempts at low carbon transitions on the other. We need, in other words, to think of low carbon transitions in terms of what Luque-Ayala et al. (Chapter 2, this volume) call an 'embedded' model of decarbonization, integral to the process of urban development. That is, at the same time as various actors and institutions are developing various plans, projects and experiments to stimulate low carbon development, they are at the same time doing other things that go in the opposite high carbon direction. As a result, normalized high carbon decisions and the cultures that underpin them are undermining their low carbon experiments. As a consequence, the political dynamics that make such normality difficult to shift ought also to attain significant attention, since such entrenched regimes need also to be shifted. Urban planning is one such site, certainly in North America. While it is the case that many cities across the continent have developed all sorts of innovations in climate policy, they nevertheless have struggled for the most part to shift the entrenched urban development dynamic centred on road and housing construction, combined with a series of contestations over urban sprawl and site-specific contestations over particular projects (see also Campbell, 1996).

Our argument is that exploring the politics of urban intensification underscores the simultaneously cultural and political-economic character of this politics. That is, we see a number of conflicts that are simultaneously about deep questions of subjectivity and identity and about the overall urban strategy and relations between key economic actors (especially property developers) and the local state. These two intertwined dynamics shape conflicts over intensification, which calls into question the meanings of what Luque-Ayala et al. (Chapter 2, this volume) refer to as 'everyday' and 'mundane' infrastructures – roads, buildings, sidewalks and the like. Alternatively, in the terms set out by Bulkeley, Paterson and Stripple (Bulkeley et al. 2016), these are conflicts where dissent against low carbon transitions is generated by desires to maintain specific types of devices in the urban space.

Exploring the conflicts over intensification thus acts as analysis of what might be called the everyday un-politics of climate change. It focuses on the sites where patterns of GHG emissions are set in train by everyday decisions by municipal governments – where they allow what kinds of buildings to be built. On the one hand, these contradictions arise out of the sorts of processes that generate the decisions to build or not build in particular places – the political economy of intensification. On the other hand, they arise out of the cultural norms and practices that underpin the widespread hostility to intensification – the cultural politics of intensification. Between them, exploring these processes enable us to understand better the politics of low carbon transitions.

The chapter explores this through an analysis of newspaper coverage of conflicts over intensification in Ottawa, Ontario. Ottawa serves as a useful example of a city that struggles to develop a response to climate change. It has developed a number of plans to address GHG emissions, none of which have become deeply institutionalized or had significant political capital invested in them. Ottawa's emissions have reduced in the last few years, but almost entirely due to the coal phase-out in Ontario (see below). And it is not the site of the significant experimentation that other scholars report in many cities (Hoffmann, 2011; Bulkeley and Castán Broto, 2013; Castán Broto and Bulkeley, 2013; Evans and Karvonen, 2014; Bulkeley et al., 2015). So it's a useful case for exploring what makes it a struggle to pursue low carbon development in many cities.

After outlining in a bit more detail the question of intensification and the reasons for focusing on it in relation to thinking about low carbon transitions, we turn to exploring a set of conflicts over specific intensifications in Ottawa. We outline the general themes that animate conflicts over these projects and then explore the questions of culture and political economy in more detail, via analysis of conflicts over three specific projects in the city, before returning to consider some broader implications for thinking about low carbon urban transitions.

2 Density, intensification and low carbon transitions

We know in general that increasing urban density is crucial to improving the overall GHG performance of a city (e.g. Hoornweg et al., 2011). Low density cities have structurally higher emissions than denser ones, primarily due to increased automobile dependence, and secondarily to increased home size and thus heating/cooling bills. European cities tend to have below half the GHG emissions per capita than North American cities as a consequence. In the Ontario context, electricity is highly decarbonized through a mix of hydroelectricity (22%) and nuclear (57%), having entirely eliminated coal-fired generation in 2014, and with rapid growth in wind and solar since the introduction of a feed-in tariff in 2009.[1]

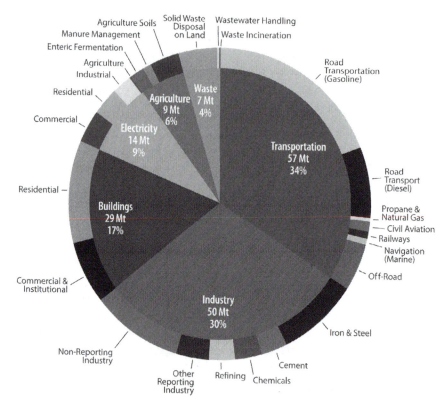

FIGURE 12.1 Ontario's 2012 greenhouse gas emissions by sector.
Note: Mt, metric ton.
Source: MECC, 2015, p. 30.

This means that transport emissions and direct energy use in buildings (natural gas mostly, with some propane in rural areas) are extremely important in Ontario's GHG emissions, accounting for 34% and 17% respectively (Figure 12.1). Therefore, central to decarbonisation is the transformation of the transport sector. And central to understanding this is understanding the causes of automobile dependence – the patterns of urban development. Low carbon transitions involve remaking the urban flows integral to life in cities.

Most North American cities have in place, going back to the 1990s, policies aimed at increasing urban density. This is increasingly articulated in relation to climate change, but was originally driven by a combination of fears about the decline of urban cores, the fiscal costs of low density sprawl, the social contestation over city form, and urban air quality concerns. These rationales still dominate official arguments for intensification. So the push for increased urban density arises in effect out of a series of internal contradictions within the political economy of urban development arising in particular out of the costs of car-dominated low density development.

Increasing urban density has been pursued, at least in the Ontario context, through the notion of 'intensification' as a planning tool. It refers, in bureaucratic language, to the percentage of new residential buildings that must be built inside the existing built-up area. Municipal councils in Ontario are required by the province to set targets in their strategic planning documents. At the same time, however, they are also required to have enough land outside that boundary zoned for residential development equivalent to their projected demand for residential housing over the next seventeen years. So the pursuit of intensification is structured within the existing urban political economy of development – a particular set of relations between the City Council, property developers and community groups, as well as relations among councillors (notably between urban and suburban councillors) enforced via the jurisdictional subordination of municipalities to the province.

Addressing climate change has only recently become stated as an explicit goal of intensification in Ontario. Intensification was first declared as a goal in the Provincial Policy Statement (the Ontario document that guides municipal planning) in 1996, and there was no mention of climate change at all – intensification had principally an economic or fiscal rationale – to reduce the costs of infrastructure provision by municipalities. In the revised statement in 2005, climate change is mentioned but only tangentially in a section on 'settlement areas', stating that they should be done so as to 'minimize negative impacts to air quality and climate change' (MMAH, 2005, sec. 1.1.3.3). Only in the 2014 statement does addressing climate change become an explicit goal guiding municipal planning, and thus all aspects of the intensification process (see MMAH, 1996, 2005, 2014). Mentions of climate change are peppered throughout the report, both regarding the implications of policy for climate change and vice versa (via the notion of resilience). Planning authorities are instructed via the 2014 policy statement to support land-use and development patterns which promote climate change mitigation and adaption. Climate change as a goal is represented in planning by compact form (intensification), active transportation, focusing travel-intensive employment and commercial activities in regions well served by transit, maximizing renewable energy opportunities, increasing vegetation where possible, and pursuing energy conservation and efficiency in the design and construction of buildings. But neither 'carbon' nor 'transitions' is ever mentioned, key framings for understanding climate change as a question of low carbon transitions. So for most of this period, studying intensification is mostly a heuristic for the sorts of conflicts that can be expected as cities engage in low carbon transitions rather than direct observations about such transitions.

3 Studying intensification in Ottawa

To explore the politics of intensification in Ottawa, we collected media reports of the public discussions of intensification.[2] We collected all news media documents from local media in Ottawa between 2001 and 2014 where the keyword 'intensification' appeared. This gave 506 articles, all of which used intensification in the context we were interested in. Of these, the vast majority reported on specific intensification projects; there were basically none that reported on intensification within the overall city planning process, and only a handful that focused in a broader way, usually via an extended interview with a key local politician or developer. Furthermore, most of the coverage focused specifically on the conflicts involved over these projects. Much of the coverage was focused on public meetings that were either formal consultation meetings organized by the city or developers, or community-organized meetings around specific projects. As such, and despite the limits of media as a source, it is a useful way to explore the character of the conflicts over intensification projects. And reflecting the point above, very few of our sources mention climate change or greenhouse gases at all; climate change is tangential to the explicit politics of intensification. To be specific, climate change appears in only six of our 506 articles, almost all simply invoking the general claim that pursuing higher urban density is important to deal with climate change, and 'carbon' only appears twice, both simply stating that intensification enables a lower 'carbon footprint' for the city. Neither climate change nor carbon are invoked explicitly in any of the specific conflicts we discuss.

We analysed these sources in a number of ways. We looked at some of the basic demographics to identify the main patterns. The projects were geo-coded and analysed using GIS software (ArcGIS) to see the basic spatial patterns (i.e. were the conflicts principally over inner-urban or suburban projects), we used it to identify some of the key actors who recur across a range of projects, and we coded the principal discursive themes, for a subset of the projects which were heavily discussed in the media reports (and thus a source of relatively rich data).

At least, as represented in media reports, the significant majority of the projects discussed were in the urban core. That is, intensification focuses in Ottawa on large projects in inner-urban wards. Figures 12.2 and 12.3 show this graphically. They show that 68% of all projects mentioned are located in one of four wards – Somerset (which includes the city centre), Capital (the ward directly to its south), Kitchissippi (to its west), and Rideau-Vanier (to its immediate east).

We turn now to exploring three elements in the narratives in these reports that we argue are key to understanding the political dynamics of intensification. For each, we develop the general narrative but then illustrate it specifically regarding conflicts over one specific construction project (for a summary description of these projects, see Table 12.1).

4 Political economy: intensification as spatial fix

The first is that the dynamics of intensification do not mark a significant break from traditional political-economic dynamics of urban growth in North American cities. These used to be widely characterized as being driven by a 'growth coalition' consisting of property developers, city politicians and bureaucrats, combined with a planning system which is restricted to zoning land according to use – agricultural,

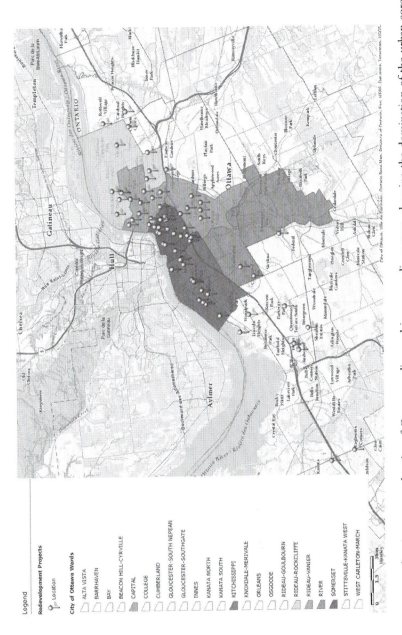

FIGURE 12.2 Location of projects across the city of Ottawa as discussed in our media sources, showing the domination of the urban core.

Note: Each green pin indicates an individual project. The grey lines are city ward boundaries. It shows clearly that the significant majority of the projects are located in the urban core. Indeed fully 68% of the projects are in the four central wards of Somerset (the city centre, with the most projects), Kitchissippi, Capital and Rideau-Vanier.

Source: map drawn using ArcGIS.

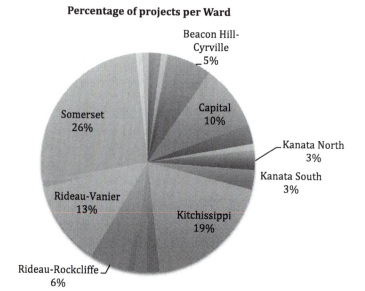

Percentage of projects per Ward

FIGURE 12.3 Distribution of projects by city ward.

commercial, residential and so on (e.g. Molotch, 1993; Dowding, 2001; this approach is relatively out of fashion in urban studies). The dynamic works such that residential construction occurs in outer urban areas because developers buy up cheap agricultural land on the urban fringe, lobby to get it rezoned as residential or commercial, thus directly realizing speculative gains in the land value, but then build on it to maximize the value from that land. Since the land they buy is cheap, there is no particular reason to build at high densities. The other side of the equation is that developers provide the principal source of campaign funding for city politicians, who (with a few exceptions) are not typically in political parties that might provide alternative sources for such finance. Politicians are thus frequently beholden to developers' interests and see the interests of developers as coterminous with those of the city as a whole (see Flyvbjerg, 1998; Jonas et al., 2011).

As various jurisdictions started to realize the contradictions of sprawl highlighted above, and developed the notion of intensification as a bureaucratic means of attempting to shift investment towards infill development and the redevelopment of urban areas at higher densities, this shift was not accompanied by shifts in the broader planning system to enable more strategic direction of investment or decide where new building should occur. That part of the system has remained organized around the purchase of land by developers, applications to build on or redevelop it – according to developers' existing business models – and pressuring councils to accept proposals for new buildings. Developers (some of them at least) responded to the intensification by identifying and purchasing land and then developing strategies for its redevelopment. Intensification thus operates in the manner of a 'spatial fix', where land zoning operates to construct

TABLE 12.1 Case studies of redevelopment projects and conflicts in Ottawa.

Case study project	City ward	Project description	Description of the conflict
Lansdowne	Capital (Ward 17)	Lansdowne Park was a plot of 40 acres of undeveloped land with remnants of a stadium and heritage buildings by the Rideau Canal (a UNESCO world heritage site) in downtown Ottawa. The City entered a public–private partnership with the OSEG to redevelop the land into a mixed-use development featuring residential, commercial and recreational uses.	Local community groups lobbied at City Hall, organized volunteers for door-to-door campaigning, and fund-raisered money to appeal to OMB and to the Ontario Superior Court of Justice. They were supported strongly by their municipal councillors.
Laurier and Friel	Rideau-Vanier (Ward 12)	An acre of land was proposed for redevelopment into a mid-rise, mixed-use building to accommodate the rapidly growing student population of the nearby university. The building would be 9 stories, 190 units, and house more than 600 students in the downtown core.	The local community association, the councillor of the ward, the mayor and most of the council opposed the project. Reasons against the project: negative consequences to the heritage quality of the neighbourhood, the height of the building and effects on the streetscape, the consequences in terms of neighbourhood diversity and character (i.e. increased student population), and effects on the liveability of the neighbourhood.
114 Richmond	Kitchissippi (Ward 15)	A 5.5 acre convent, established in 1910, was purchased and a development plan was proposed along Wellington St West (an area where intensification was being promoted by the City). Of the three buildings proposed, two were to be built along the main street with heights of 11 and 8 stories tall where zoning deemed 6 stories to be the maximum height allowance.	The convent site had not received official heritage designation, which resulted in the mobilization of the community to try to secure designation and public access to the site. Community associations appealed the approval of the project plan and entered into negotiations with the project proponent to alter the design plan.

Note: OMB, Ontario Municipal Board; OSEG, Ontario Sports and Entertainment Group; UNESCO, United Nations Educational, Scientific and Cultural Organization.

market value for developers and create incentives for investment and a cycle of accumulation (cf. While et al., 2004).

In Ottawa (at least) this has meant a preponderance of high-rise apartment buildings, where existing areas of unbuilt, brownfield or low-rise buildings were converted into high-rise buildings. In some instances, these have been shaped by Community Design Plans, drawn up by the Council in consultation with local community organizations, that make specific recommendations regarding energy efficiency, green roofs, solar energy and similar specifically low carbon technologies, but these are never more than guidelines and are tangential to the main drivers of intensification. Frequently, developers applied to build buildings considerably higher than those specified in existing zoning requirements (provoking conflicts, see below), in order to maximize returns on their investment in the land and construction. According to Bill Malhotra, founder of Claridge Homes (whose projects feature heavily in our data), intensification produced a sort of land speculation centred on the premium to be gained from projects meeting intensification guidelines. Projects that get constructed are determined by developers' judgment of profit potential, the higher the better for them (Tencer, 2004; see also Langston, 2011). This has been intensified by the Council's strategy to focus intensification near key transit nodes, around stations in Ottawa's Transitway and O-Train (and especially near the light rail line, under construction as of 2017).[3] This is in line with planning notions of building along dense corridors, but has created intense spatial fix dynamics in particular parts of the cities near key transit nodes in gentrifying areas. Projects in Kitchissippi Ward, along Wellington Street West (near Westboro station on the Transitway), as well as at 500 Preston (by an O-Train station) and near the Bayview Transitway station, have all emerged rapidly, many of which have been highly contested.

Related, the form of the appeal system concerning development decisions similarly favours the strategies of developers. In the Ontario context, this is the Ontario Municipal Board (OMB) and, for the dynamics of sprawl, developers could normally rely on OMB to overturn Council decisions should the Council turn down a specific development. While OMB would formally be required to take intensification goals into account, the general consensus is that OMB largely favours developers in their appeals. The same became true once intensification became a policy tool, although this time the appeals are not over being allowed to build on land zoned as agricultural or against the amount of land on the urban fringe zoned for residential development, but over questions of the height of buildings, setback from the road, closeness to a property line, and the like. This ability to go to OMB then structures debates within Council, since the Council would have to incur legal costs to go to OMB (which even if it would win sometimes it may judge not worth the cost).

Among opponents of intensification projects, this dynamic is reflected in what can be characterized as a 'developers vs community' discourse. This is particularly prominent in the discourse of Clive Doucet, city councillor for Capital Ward 2001–10 (and prior to that councillor for the ward in the previous Ottawa–Carleton

Regional Council[4]), but is also frequently invoked by other councillors in the urban core, notably Diane Holmes (Somerset Ward), Tobi Nussbaum (Rideau-Rockliffe, to the north-east of Rideau-Vanier) and David Chernushenko (Capital Ward from 2010 onwards). This argument is that, in Chernushenko's words, "private interests are determining the speed and character of urban intensification" (as quoted in Pilieci, 2014). In effect this is a left-populist opposition to intensification. At times it is connected to arguments that councillors are influenced by the money they get from developers, with one councillor, Tobi Nussbaum, proposing to ban corporate funding of municipal candidates in order to prevent this sort of influence.[5]

The project at Lansdowne park exemplifies this sort of dynamic most effectively (Figure 12.4). Located next to the Rideau Canal in the downtown Glebe neighbourhood of Ottawa resides Lansdowne Park. The park, until construction began in 2012, was a plot of 40 acres of undeveloped land with remnants of a stadium and a few heritage buildings. The City of Ottawa was unable to invest in the land given budget constraints and the fact that any funding secured through provincial or federal channels would be allocated to higher priority areas (such as transit and affordable housing). The City entered into a public–private partnership with the Ontario Sports and Entertainment Group (OSEG) in order to redevelop the piece of land into a mixed-use development featuring residential, commercial and recreational uses. The contestation between the community and the city/developers highlights this narrative as the planning process was largely ignored in favour of the developers' and city's goals.

The City of Ottawa and OSEG partnership is likely the best scenario that could have unfolded for the previously paved, vacant space of Lansdowne. However, there was still an outburst of opposition from the community which led to what was

FIGURE 12.4 Redevelopment of Lansdowne Park.
Source: Xavier P.-Laberge, image used with permission.

described by one journalist as "the opening shot of a war in which residents will fight to preserve the quality of their neighbourhoods" (Gray, 2011). The conflicts that arose as a result of the Lansdowne redevelopment project can be attributed partially to the demographic of the surrounding neighbourhoods. These neighbourhoods are affluent and contain a large number of public sector professionals (government, universities, hospitals); data from 2010 revealed the neighbourhoods as the most politically engaged in the country with reference to their political donations (Ottawa Citizen, 2010). The mobilizing power of the neighbourhoods was extensive; Friends of Lansdowne was able to undertake exhaustive lobbying at City Hall, organize volunteers for door-to-door campaigning, and raise enough money to not only appeal to OMB but also to the Ontario Superior Court of Justice. They were supported strongly by their successive councillors, Doucet and then Chernushenko.

It is difficult to say whether the Friends of Lansdowne would have just as fiercely opposed the development plan had it been accepted and implemented through normal planning and consultation processes. There were multiple design aspects of the original site plan that were changed throughout the appeal process and many residents attribute the end result as a partial win for the community group. However, it is clear that the City's decision to pursue a public–private partnership and their failure to host a sufficient consultation process with the community was a rallying point for those opposed to intensification and provided support for the 'developers vs community' narrative.

The City of Ottawa had started off with a design competition accompanied by a public consultation website which had been created in order to gather the necessary information to issue a Request for Proposal. This process was quickly suspended after the City received the proposal for the redevelopment of Lansdowne from OSEG and a recommendation report from Council that favoured the pursuit of the OSEG partnership (Friends of Lansdowne Inc v. Ottawa (City), 2011). The skipping of a competitive bidding process and apparent disregard of public consultation led to the distrust of the developers and the city, with one newspaper article noting that "residents, many of whom are property owners, wonder if the development community runs this municipality and if the city, overruled by the OMB, can do anything about it" (Gray, 2011).

Given the importance of intensification in creating the structural conditions for low carbon transitions, it is important to understand the underlying political-economic drivers and the conflicts that surround this process. The explicit mobilization of low carbon transitions, or more simply, the mention of climate change and GHG emissions is notably absent from the public discourse promoting intensification in Ottawa. Rather, the process of intensification is driven by the changes to the provincial planning rules that reshape incentives for developers.

5 Competing narratives of intensification

The 'developers vs community' narrative underscores that the politics of intensification is less about whether or not to intensify, but rather about how to do it:

intensification is not one, but could be various paths to low carbon transition (see Luque-Ayala et al., Chapter 2, this volume). The second theme in the news stories is that the conflicts over specific intensification projects are not usefully understood as conflicts between proponents and opponents of intensification per se. Rather, they are conflicts between competing narratives of intensification, a tension that students of urbanism might recognize readily as 'Le Corbusier vs Jacobs'. On the one hand there is a vision of a densely built hyper-modernist form of development centred on high-rise buildings and car-oriented development; on the other one, of dense, mixed-use, 'organically' developing neighbourhoods centred on walking, cycling and transit (e.g. Jacobs, 1961; Berman, 1982).

These competing narratives reflect the political economy sketched above. The opponents of specific projects at times couch their opposition to high-rise intensification explicitly in terms of the approach to cities of Jane Jacobs,[6] notably Clive Doucet and David Chernushenko. Doucet routinely framed his opposition to high-rise buildings in these terms. There is recurrent mention of Paris not having buildings above eight stories yet being highly dense, of the importance of walkable neighbourhoods, of the roads they are on becoming 'traffic sewers', of the argument that skyscrapers in fact feed sprawl. On the last of these themes, this is stated to be because no one in practice walks from them; instead they simply take the lift to the underground car park from where they drive. Moreover they are family unfriendly so feed the process of families moving to the suburbs, leading to a crisis of urban schools, which then becomes a self-feeding process. In Doucet's words, "each decision will take us in one direction or the other: toward creating a city based on vibrant, mid-rise urban villages connected primarily by rapid rail and rapid bus, or toward more of the same – more highrises downtown, more low-density, distant suburbs, and more traffic sewers to connect them to the malls and downtown highrises" (Doucet, 2004).

The narrative of proponents is never explicitly expressed in terms of Le Corbusier or his various disciples, but nevertheless the aesthetics associated with high-rise development, combined with the technocentric vision of urbanism associated with this, is clearly present (see below on height, for example). Doucet in particular is framed by opponents as suffering from simple nostalgia (Denley, 2010), reinforcing this association of high-rise intensification as modernist.

The conflict that arose when Viner Assets proposed a redevelopment project at Laurier and Friel in the downtown neighbourhood of Sandy Hill clearly demonstrates the disconnect between how community members/council representatives and how planners/developers feel intensification should be performed. The site itself, consisting of almost an acre of land and four existing buildings, was slated for redevelopment into a mid-rise, mixed-use building to accommodate the rapidly growing student population of the nearby University of Ottawa. The building would be 9 stories tall and consist of 190 units, providing housing for more than 600 students in the downtown core within walking distance to campus, amenities and nearby transit stations. In a neighbourhood that is experiencing a housing crunch due to property owners rapidly converting their dwellings into student rental apartments, one would expect residents to be pleased with the proposal. The introduction of more student-focused housing contained to one building managed by a group that had listed 24-hour security and maintenance staff as part of the proposal should reduce both the conversion of properties and the noise complaints in the area.

Conversely, the local community association, the councillor of the ward, the mayor and the majority of council opposed the project. They felt that the building plan was 'too intense' for the area and voted against the proposal when it reached council in spite of the planning committee recommending the project for approval. This was a unique situation; it is rare that council goes against the planning committee's recommendations for redevelopment projects. The decision, and the subsequent overturn of that decision by OMB, highlighted that there is a distinct difference in how each group views intensification in practice, with the community leaning towards an idealized version of Jacobs' liveable streets.

The analysis of the articles on the project revealed that residents were concerned about the heritage quality of the neighbourhood, the height of the building and effects on the streetscape, the consequences in terms of neighbourhood diversity and character (i.e. increased student population), and the liveability of the neighbourhood (e.g. Chianello, 2014b). The community association and the council were clearly against the development because it was for students, suggesting that students should be redistributed across the city using transit-oriented development. When this argument was met by criticism from members on the planning council as 'blatant ageism', the community transitioned to an argument based on heritage and the liveability of the neighbourhood (Woods, 2014).

Most of the area surrounding the site was low-rise buildings and the area was zoned as a low-rise neighbourhood under the City of Ottawa's secondary plan for Sandy Hill. This zoning document, which was twenty years old at the time, did not reflect the intensification strategies that the city had since adopted. The city manager of development review testified at the OMB hearing in favour of the development, stating that the neighbourhood plan was outdated (Viner Assets Inc. v. City of Ottawa, 2015). And while some buildings nearby are heritage designated and the site itself was on the heritage reference list (but not officially designated or protected), it was determined that the project would not adversely affect heritage buildings. Ruling in favour of the developer, OMB signalled that this redevelopment project is in line with both the municipal and provincial statements on intensification and density targets.

To address the disconnect between what planners/developers and the community/their council members see as 'good practice' for intensification, the City of Ottawa has begun to manage expectations using city planning documents such as the neighbourhood secondary plans and zoning documents. Community Design Plans have emerged as non-statutory documents that act as guidelines for increasing intensification while maintaining compatibility with existing neighbourhoods (Ottawa City Council, n.d.). This consultation process allows residents the opportunity to give input into the physical attributes of future developments such as setback from street, construction materials, green roofs, solar panels, drip irrigation, permeable site surfaces and stipulations for greenspace.

6 Heritage and height: the 'character of the neighbourhood'

These conflicts over competing narratives of intensification frequently played out in the language of the 'character of the neighbourhood'. This operates as a cipher for opposition to intensification projects (while not appearing to oppose intensification per se). Projects are often opposed on the grounds of 'appropriateness' for the existing neighbourhood. In the Laurier/Friel project discussed above, this was usually couched, explicitly or implicitly, in terms of a hostility to the neighbourhood becoming more dominated by students. The Mayor voted against the proposal saying "would be wise to send a message that this is an important heritage community that's under stress" (Mueller, 2014). Another opponent stated that "We will witness a moral decay of society and community living," a thinly veiled reference to worries about student housing (Woods, 2014). In other projects, anxieties about social class are associated directly with questions of 'appropriateness' (e.g. Stern, 2010).

Beyond this reference to the specific social character of the neighbourhood, the two most significant bywords through which this unease is expressed are in terms of height and heritage. Conflicts over the number of stories a building can be built to too often form the basis for much of the antagonism felt by the community. Developers either apply for building permits to build higher than existing zoning allowances, provoking conflicts at that stage of the process, or then seek after a building has been approved to build higher than the originally proposed height. Either way they are seeking to increase the profitability of the project. Opponents emphasize the problems that the extra height causes, in terms of the neighbourhood being disturbed aesthetically by the presence of high buildings, or also in terms of material effects – shadowing of other buildings, extra parking problems because of the increased number of apartments, and so on (e.g. Brady, 2014). Conversely, developers at times explicitly promoted their buildings in terms of height: in the words of Shawn Malhotra of Claridge Homes: "point towers traditionally are beautiful buildings. When you have short, stumpy buildings there's only so much you can do" (quoted in Langston, 2013).

The heritage question is alluded to above regarding the Laurier/Friel project. But it is perhaps most readily explored in relation to the project at 114 Richmond Road (in Kitchissippi Ward). This concerned a particular heritage building – an old convent, La Soeurs de la Visitation. This convent, established in 1910, was put up for sale in 2009. The convent was on a substantial 5.5 acre plot of land, most of which was undeveloped, along Wellington Street West in an area where intensification was being promoted. The site was also two blocks west of a very significant conflict between a community association and developers over the construction of a large supermarket in the early 2000s.

Similarly to the Laurier/Friel case, the convent site had been listed on the heritage reference list but had not yet received official designation or protection. After the piece of land had been put on the market, the City of Ottawa quickly lobbied for the building to be declared a heritage building which would limit the type of development that could occur around it. Before the heritage designation process was complete, the parcel was purchased by Ashcroft.

Ashcroft released a design plan that kept pieces of the convent intact but doubled the height allowed in the neighbourhood's secondary plan. Unlike Laurier and Friel, city-wide consultations two years earlier had resulted in an updated plan for Wellington West that deemed six stories to be the maximum allowance on the main street. The first proposal set out by Ashcroft suggested a three building redevelopment project with two buildings fronting onto the main street with eleven and eight stories. The design also outlined plans to preserve the convent and create walkways between the buildings on the main street that would enter into the convent grounds, allowing the space to be accessed by the public.

The result of the unveiling of Ashcroft's plan was the mobilization of the community in opposition. A heritage designation for the site had still not been secured and the local community groups had responded by creating petitions and lawn sign campaigns, writing in to local newspapers, and calling for public meetings with their councillor. The city explored the idea of imposing a levy-tax to acquire a piece of the land and continued to delay the project using various legal techniques until a partial heritage designation was secured. The plan was eventually approved by the City and appealed by three local community groups at OMB; four months of negotiations resulted in the appeal being dropped. The project has since been constructed, with access to the convent maintained and a final height of nine stories on the frontage of the main street.

7 Conclusions: ambivalence, intensification and low carbon transitions

The observations above suggest and support the claim that the political conflicts associated with low carbon transitions are expressed via a fundamental ambivalence. We should underscore that in none of the conflicts we analysed did climate change figure explicitly in the discourse of either proponents or opponents of projects (at least as represented in news media coverage): our analysis is thus of the sub-politics of low carbon transitions. But if we take the competing narratives around intensification, there is a paradox in that those political actors who are most in favour of intensification *in general* (on the grounds of promoting both 'liveable cities' and addressing climate change – see Doucet, 2007, for example) are simultaneously the most likely to be the most opposed to *specific* intensification projects (and never mention climate change in those circumstances). This is the case for most of the city centre councillors such as Clive Doucet, David Chernushenko and Diane Holmes. Opposition to intensification per se was relatively rarely expressed. This is also underscored by considering the location of these conflicts within the city. The neighbourhoods where there has been most conflict are relatively mixed neighbourhoods but with a high percentage of public sector professionals. The electoral district that represents these neighbourhoods at the highest (federal) level (Ottawa Centre) has for most of this period voted in a Member of Parliament (MP) from the New Democratic Party (the centre-left social democratic party in Canada), Ed Broadbent and then Paul Dewar, the latter in particular having made climate change action a significant part of his political profile. More recently, in 2015, Ottawa Centre voted for a Liberal MP, Catherine McKenna, who immediately on election became Minister for Environment and Climate Change in the Liberal government. By contrast much of the rest of the city votes for Liberal or Conservative MPs, most of whom have not made climate change as prominent a part of their public profile as either Paul Dewar or Catherine McKenna. So this is the part of the city where action on climate change is given most attention. For example, Community Design Plans in Ottawa Centre are more likely to cite energy efficiency technologies and sustainable architectural innovations.[7]

This paradoxical observation is reflected widely in the narratives around intensification projects, exemplified in the discourse of Councillor Rick Chiarelli regarding various projects. Chiarelli states that he is in favour of intensification, but "said he doesn't want to see the city go overboard" (as paraphrased in Johnson, 2011). "Extremism will be the death of intensification. And council needs to support moderate intensification and reject the temptation to accept overkill" (Chiarelli, quoted in Johnson, 2011). This is perhaps best expressed in the commentary by Joanne Chianello, prominent local journalist:

> No one has to shed a tear for the plight of developers trying to sell homes, but it does make you wonder about the double-standard so many of us seem to have when it comes to infill.
>
> Some residents are already complaining about backyard parking because they don't want to look at cars when they sit on their decks. They also don't want cars parking on the street because that inhibits snow removal (and where would their guests park?). They absolutely don't want to see front yards taken over by parking. Yet they claim to support intensification.
>
> (Chianello, 2014a)

It is easy to see this as a question of hypocrisy, or of a simple Nimby politics, but it is better understood as a deeper process of desiring mutually incompatible goals, and thus a complicated messy politics of the reconstruction of subjectivity entailed in low carbon transitions. People like the idea of walkable neighbourhoods, great transit provision, services close to home, as well as the idea of being a low carbon city, but at the same time have deep attachments to large single-family homes, having more than one car (and thus the space to store them), and the like, as well as the specific sorts of urban environment and the history it embodies and the meanings that history gives to daily lives.

Much political action aims at closing down this ambivalence. Developers and the political right seek to limit the destabilizing effects of ambivalence, either by articulating intensification as a continuation of existing 'normal' patterns of development (as done mostly by the developers in our research) or by resolutely resisting the normative value of intensification itself (as in the case of some 'backlash' political statements by those on the right). Opponents tend to elide the ambivalence by creating the simpler 'developers vs community' narrative.

If our analysis has any purchase, it is in suggesting that a better starting point is to recognize this ambivalence and use it to open up space for asking what sorts of subjects we want to become in low carbon urban transitions, and what sorts of urban political economies might better enable these subjectivities to emerge? As Luque-Ayala et al. state (Chapter 2, this volume, p. 16), "the existing context of an intervention shapes its possibilities." In our case, the relevant contexts are the pre-existing planning process centred on passive zoning and active private developers, the other drivers for intensification as a project of increasing urban density, and the set of attachments and desires that generate resistance to specific intensification projects. One implication of Luque-Ayala et al.'s overview is that innovation depends in part on cities shifting the planning process in general to enable 'strategic niche management' – in the case of Ottawa, articulating urban intensification with explicit low carbon experiments in ways that might go beyond the simple implied effect of reduced emissions from higher densities. Cities that are widely regarded as pioneers on climate change, such as Copenhagen or Vancouver, express this sort of entrepreneurial role for the city perhaps better than Ottawa, but this case shows how at least one set of political dilemmas in low carbon urbanism lie in the tension between a private-led model of property development that shapes how intensification occurs, and generates specific sorts of political resistance even from those who profess to be in favour of intensification. A more effective pursuit of urban intensification by the city will thus need to work more strategically and actively with developers, while bringing them into participative arrangements with neighbourhoods within the city in order to mobilize the expressed desire for urban intensification and enable these sorts of affective ambivalences to be navigated explicitly rather than simply generating opposition to development. In other words, the competing visions of the future contained in intensification's politics – between a high-modernist vision centred on high-rise development, and a 'liveable streets' vision centred on walkable mixed-use neighbourhoods and mid-rise buildings – need to be articulated explicitly as different potential pathways to low carbon urbanism, and the choices entailed in each form opened up to generate potential for more strategic and experimental interventions than those generated from the current form of urban planning in cities like Ottawa.

Our hunch is that many more cities are like Ottawa than are like the pioneer cities that are often the focus of research on the urban politics of climate change. That is, they are more likely to be ones where climate change is only rarely mentioned in the official public life of the city, and it is even more rarely connected explicitly to aspects of urban life that dominate much of city politics – especially the intertwined politics of city planning, building approval decisions, and transport strategy. To this extent, the analysis thus raises two important sorts of question for future work on these types of non-pioneer cities. First, we could ask whether or not the explicit politics of climate change and/or carbon makes a difference to these conflicts over urban development, i.e. if a city frames intensification projects explicitly as low carbon initiatives, does this make them easier to sell politically? And does an explicit low carbon framing of these projects make a difference to what sort of low carbon urban transitions are favoured, or does it simply put a low carbon veneer on the sort of developer-centred initiatives we have seen in Ottawa, with their attendant problems? But second, we could ask what it might take for cities like Ottawa to shift to a more strategic mode of orientation, to think explicitly and systematically about the planning process in terms of a low carbon transition? Is it simply a question of specific political agents or are there other dynamics that favour this sort of shift in urban politics?

Notes

1 This is also the case for most of Canada, although the US is rather different, with still considerable amounts of coal-fired electricity. Ontario does still use a little coal-generated electricity given electricity imports from the US.
2 There are obviously limits and biases in media reports, as in any data source, specifically in the way that they privilege formal political actors more than others, and may have specific editorial agendas that are pursued in the stories.
3 Ottawa has had, since 1983, a dedicated bus only roadway, known as the Transitway. It has one principal line running east–west across the city, and two lines running south from this to other suburbs. There has also been one line of light rail operating, known as the O-Train, north–south from just West of the downtown core. A second line is currently being built, running east–west along the current Transitway line, but running under the city centre in a tunnel. This is due to open in 2018.
4 Ontario amalgamated a set of municipalities in 2001 to create a new, fused, Ottawa City Council. Part of the aim here was to enable more effective planning of transit and other services across the entire urban area (apart from the parts on the Quebec side of the Ottawa River) and also specifically to avoid a specific dynamic favouring sprawl – competition between different municipalities to attract expanded tax revenue via enabling residential development.
5 His campaign was unsuccessful, but the province did then ban corporate donations for municipal elections as of 2018. See Chianello 2016.
6 Jacobs has a particular resonance among Canadian urbanist political actors, especially on the political left (like Doucet and Holmes in the Ottawa case), in part because she lived much of her life in Toronto and in particular was highly involved in the successful campaigns over the 'Spadina expressway' during the late 1960s through to its cancellation in 1971. See for example Zielinski and Laird 1995.
7 This claim is based on reading the Community Design Plans across the city (Ottawa City Council, n.d.).

References

Ala-Mantila, S., Heinone, J. and Junnila, S. (2013) Greenhouse gas implications of urban sprawl in the Helsinki Metropolitan Area. *Sustainability*, 5(10), 4461–4478.

Bart, I. (2010) Urban sprawl and climate change: a statistical exploration of cause and effect, with policy options for the EU. *Land Use Policy*, 27(2), 283–292.

Berman, M. (1982) *All That Is Solid Melts into Air: The Experience of Modernity*. London: Verso.

Brady, S. (2014) The fight for height; finding a balance between the needs of community, city and developer when it comes to intensification is not easy. *Ottawa Citizen*, 14 June, p. G1 (online at Factiva).

Bulkeley, H. and Castán Broto, V. (2013) Government by experiment? Global cities and the governing of climate change. *Transactions of the Institute of British Geographers*, 38(3), 361–375.

Bulkeley, H., Castán Broto, V. and Edwards, G. (2015) *An Urban Politics of Climate Change: Experimentation and the Governing of Socio-technical Transitions*. London: Routledge.

Bulkeley, H., Paterson, M. and Stripple, J. (2016) *Towards a Cultural Politics of Climate Change: Devices, Desires and Dissent*. Cambridge: Cambridge University Press.

Campbell, S. (1996) Green cities, growing cities, just cities?: urban planning and the contradictions of sustainable development. *Journal of the American Planning Association*, 62(3), 296–312.

Castán Broto, V. and Bulkeley, H. (2013) A survey of urban climate change experiments in 100 cities. *Global Environmental Change*, 23(1), 92–102.

Chianello, J. (2014a) Infill rules not easily met by all; conformity can be challenging. *Ottawa Citizen*, 20 March, p. C1 (online at Factiva).

Chianello, J. (2014b) Watson likely looking toward next election. *Ottawa Citizen*, 27 March, p. C1 (online at Factiva).

Chianello, J. (2016) Ottawa councillors unrattled by ban on corporate campaign contributions. CBC News online, 5 July. Available at: http://www.cbc.ca/news/canada/ottawa/ottawa-councillor-corporate-campaign-funding-ban-1.3638660 (accessed 5 January 2017).

Denley, R. (2010) Pity Doucet isn't our needed antidote to the 'twins'; a crazy quilt of contradictions. *Ottawa Citizen*, 9 October, p. C1 (online at Factiva).

Doucet, C. (2004) Two storeys matter in a livable city. *Ottawa Citizen*, 14 January, p. B4 (online at Factiva).

Doucet, C. (2007) *Urban Meltdown: Cities, Climate Change, and Politics-as-Usual*. Gabriola Island, BC: New Society Publishers.

Dowding, K. (2001) Explaining urban regimes. *International Journal of Urban and Regional Research*, 25(1), 7–19.

Evans, J. and Karvonen, A. (2014) 'Give me a laboratory and I will lower your carbon footprint!' – urban laboratories and the governance of low-carbon futures. *International Journal of Urban and Regional Research*, 38(2), 413–430.

Flyvbjerg, B. (1998) *Rationality and Power: Democracy in Practice*. Chicago: University of Chicago Press.

Friends of Lansdowne Inc. v. Ottawa (City) (2011) ONSC 4402 10–49352.

Glaeser, E. and Kahn, E. (2010) The greenness of cities: carbon dioxide emissions and urban development. *Journal of Urban Economics*, 67(3), 404–418.

Gray, K. (2011) Lessons from Lansdowne. *Ottawa Citizen*, 3 August, p. A11 (online at Factiva).

Hoffmann, M. J. (2011) *Climate Governance at the Crossroads: Experimenting with a Global Response after Kyoto*. New York: Oxford University Press.

Hoornweg, D., Sugar, L. and Trejos Gómez, C. (2011) Cities and greenhouse gas emissions: moving forward. *Environment and Urbanization*, 20(10), 1–21.

Jacobs, J. (1961) *The Death and Life of Great American Cities*. New York: Vintage.

Johnson, J. (2011) Residents oppose Nepean condo development: intensification project in line with city mandate. *Ottawa Citizen*, 12 August, p. C5 (online at Factiva).

Jonas, A.E.G., Gibbs, D. and While, A. (2011) The new urban politics as a politics of carbon control. *Urban Studies*, 48(12), 2537–2554.

Langston, P. (2011) Vanier: a new view; slowly moving away from its troubled past, the neighbourhood is attracting condos, infill housing and a stronger sense of community. *Ottawa Citizen*, 6 August, p. I1 (online at Factiva).

Langston, P. (2013) 'A milestone of urbanization'; the skyline around Carling and Preston is about to change rapidly. *Ottawa Citizen*, 11 May, p. I4 (online at Factiva).

MECC (Ontario Ministry of the Environment and Climate Change) (2015) *Ontario's Climate Change: Discussion Paper 2015*. Toronto: MECC.

MMAH (Ontario Ministry of Municipal Affairs and Housing) (1996) *Provincial Policy Statement 1996 (amended in 1997)*. Toronto: MMAH.

MMAH (Ontario Ministry of Municipal Affairs and Housing) (2005) *Provincial Policy Statement 2005*. Toronto: MMAH.

MMAH (Ontario Ministry of Municipal Affairs and Housing) (2014) *Provincial Policy Statement 2014*. Toronto: MMAH.

Molotch, H. (1993) The political economy of growth machines. *Journal of Urban Affairs*, 15(1), 29–53.

Mueller, L. (2014) Sandy Hill student resident plan rejected by council; appeal to OMB likely after proposal for 180-unit apartment turned down. *Ottawa West News*, 3 April. Available at: https://www.ottawacommunitynews.com/news-story/4446662-sandy-hill-student-residence-plan-rejected-by-council-appeal-to-omb-likely-after-proposal-for-180-unit-apartment-turned-down/ (accessed 19 January 2017).

Newman, P.W.G. and Kenworthy, J.R. (1989) *Cities and Automobile Dependence: A Sourcebook*. Brookfield: Gower Technical.

Ottawa Citizen (2010) Political animals (editorial). *Ottawa Citizen*, 12 May, p. A12 (online at Factiva).

Ottawa City Council (n.d.) Community design plans. Available at: http://ottawa.ca/en/city-hall/planning-and-development/community-plans-and-design-guidelines/community-plans-and-0#what (accessed 27 January 2017).

Pilieci, V. (2014) Planning is still a big civic issue: two candidates say rising housing costs are also a concern. *Ottawa Citizen*, 8 October, p. A6 (online at Factiva).

Stern, L. (2010) Growing intense. *Ottawa Citizen*, 30 October, p. B6 (online at Factiva).

Tencer, D. (2004) Breathing life into downtown. *Ottawa Citizen*, 21 October, p. C3 (online at Factiva).

VandeWeghe, K. and Kennedy, C. (2007) A spatial analysis of residential greenhouse gas emissions in the Toronto Census Metropolitan Area. *Journal of Industrial Ecology*, 11(2), 133–144.

Viner Assets Inc. v. City of Ottawa (2015) Toronto: Ontario Municipal Board, PL140348. Available at: https://heritageottawa.org/sites/default/files/OMB_DECISION%20pl140348-jan-27-2015.pdf (accessed 1 August 2017).

While, A., Jonas, A.E.G. and Gibbs, D. (2004) The environment and the entrepreneurial city: searching for the urban 'sustainability;fix' in Manchester and Leeds. *International Journal of Urban and Regional Research*, 28(3), 549–569.

Woods, M. (2014) Neighbours fail to halt student housing: 'we will witness a moral decay of society and community living', opponent says. *Ottawa Citizen*, 26 February, p. C3 (online at Factiva).

13

POST-DEVELOPMENT CARBON

Andrés Luque-Ayala

> Stemming from the combined crises of food, energy, climate, and poverty, these
> transition discourses – particularly prominent in the areas of ecology, culture and
> spirituality – can be seen as markers for postdevelopment, or as challenges to modernity
> more generally.
>
> Arturo Escobar, Encountering Development *(preface to the 2012 edition, p. viii)*

1 Introduction

What might low carbon urbanism look like in a post-development world? This
chapter sets out a research agenda around the mobilization of low carbon urbanism
outside dominant understandings of development. It revisits the post-development
debate within development studies, to reflect on its possible contribution to the
low carbon city and its politics. Here, low carbon urbanism, understood as both a
set of narratives and their manifestation in a series of interventions, finds a space
within post-development and its critique of development as a discursive formation
with political aims, its denaturalization of development as the only possible path for
Asia, Africa and Latin America, and its examination and encouragement of narratives and representations that transcend development (Escobar, 1995). The chapter
asks questions about the potential, limitations and implications of addressing climate
change in cities in the global South through initiatives and social movements
which operate via what could be called a post-development logic – that is, taking
steps towards displacing liberal ontologies and capitalist imaginaries from their
central role as organizing principles of both economy and society. It proposes the
idea of *post-development carbon* as a way of understanding (i.e. an analytical position)
and advancing (i.e. a narrative or intervention) low carbon initiatives from three
interlinked positions. First, as templates for imagining and enacting a post-fossil fuel
world. Second, as steps towards a post-capitalist economy, seen as an economy

made of diverse capitalist and non-capitalist practices. And third, as strides towards a post-liberal world, where social life is not primarily determined by the traditional pillars of modern European-Western rationality, such as the economy, private property or an ontology that favours the individual over community. Post-development carbon, as both a way of thinking and acting, opens possibilities to decentre issues of climate mitigation and a reduction in greenhouse gas (GHG) emissions from the very heart of the low carbon initiative. It also acts as a marker to examine up to what extent climate responses in the city embrace, challenge or transform dominant development modes.

Post-development carbon seeks to interrogate the possibility of "alternatives to development" imagined and enacted through low carbon and post-fossil fuel interventions. The chapter draws on three recent advances within post-development thinking. First, a recognition of a broad range of transition discourses as markers for post-development (Escobar, 2012). Second, a foregrounding of a language of hope and possibility (McGregor, 2009). And third, a political project around a transformation in how the economy is represented (Gibson-Graham, 2006 [1996]). Building on this, it seeks to open space for repositioning climate responses in the city as interventions that "liberat[e] oppressed imaginaries and alternative futures by delegitimising development, marginalising traditional development institutions and creating new spaces where unconventional pathways can be explored and enabled" (McGregor, 2009, p. 1695). Following from a brief reference to how climate mitigation is framed within development, the chapter uses a post-development approach to unpack the multiple and contradicted ways in which low carbon urbanism is being mobilized in the global South. The resulting research agenda asks whether, and how, urban low carbon initiatives are challenging traditional notions of development and devising alternative ways of imagining collective futures. Post-development thinking allows for a plural understanding of low carbon urbanism – as multiple, diverse, non-linear and at times conflicted. Such plurality is found in different vernacular expressions that might not directly foreground climate mitigation, yet embrace a transition to a post-fossil fuel world while keeping close tabs on the political implications of such changes (c.f. Mitchell, 2011; Shiva, 2008). In doing this, the chapter seeks to advance an understanding of low carbon urbanism as a matter of development modes (Luque-Ayala et al., Chapter 2, this volume). While the primary purpose of the chapter is to set the foundations for a practical and analytical framing of low carbon urbanism in cities in Asia, Africa and Latin America, the potential of the proposed ideas is not exclusive to such geographies.

Around the world, a small number of community initiatives and projects conceived in ways that are consistent with post-development carbon are mobilizing narratives and interventions in line with the objectives of low carbon and energy transitions. For example, in Enkanini, an informal settlement disconnected from the electricity grid in the outskirts of Stellenbosch, South Africa, the iShack project is testing the practicalities of living in a post-fossil fuel world while advancing a different way of conceiving modernity (see Figure 9.1 in Chapter 9, this volume). The project – a partnership between a local dwellers association and the Sustainability Institute (a non-profit trust working in partnership with the University of Stellenbosch) – is providing electricity access to residents via a solar panel kit connected to energy-efficient power lights, a television and other small appliances. For those involved, this "off-grid utility" run by a social enterprise (Sustainability Institute, n.d.) was developed as an alternative to the 'trust-and-wait' logic of service provision common in informal settlements across the global South (Swilling, 2014) – arguably, as a response to the delayed arrival of traditional development. It is important to note that climate change mitigation is not what matters for project beneficiaries, while project coordinators are quick to point out that it would be wrong to portray the project in such light. Instead, they speak of the need to change people's minds so that the (carbon-intensive) electricity grid is not seen as the only way for having energy at home. With the project originally conceived as a contribution to 'sustainable development', neither the need nor the desire for development has been eliminated. Rather, traditional understandings of development have been temporarily displaced or decentred through a narrative that foregrounds local agency, community economies and collective rights to resources such as energy despite not being within reach of the infrastructures of development (the electricity grid). Similar initiatives that transcend low carbon logics and a focus on GHG reduction while proposing a different mode of (urban) development are also present in cities in the global North, where ideas of low carbon are joined by notions of economic justice, self-governance, mutualism, affordability, equality and others (e.g. see Chatterton, 2013, and Hollands, 2015, for the case of LILAC – Low Impact Living Affordable Community – a housing project in Leeds, UK).

The specific case study used in this chapter to discuss and illustrate a post-development carbon approach is from São Paulo, Brazil. Specifically, the chapter discusses the workings of Sociedade do Sol (Society of the Sun), a non-profit involved in designing and promoting low-cost do-it-yourself solar hot-water systems (or DIY SHW systems). These systems, designed initially with the aim of providing cheap energy for the poor, are a vehicle for a range of discussions with local stakeholders on issues of environmental sustainability and climate change. Critically, Sociedade do Sol mobilizes low carbon energy towards a set of social and economic agendas that transcend concerns around climate mitigation, advocating for non-capitalist practices of service provision. The case study is illustrated with material collected via interviews conducted in São Paulo with a range of stakeholders participating in the activities of Sociedade do Sol. The rest of the chapter is structured in four parts, as follows. Section 2 introduces the critique to development elaborated by post-development scholars, and reflects on the ways in which post-development provides analytical tools for a different reading of the low carbon urbanism–development interface. Here, post-development carbon is seen as an *analytical position*. Section 3 provides a summary of the dominant paradigms by which narratives around climate change are incorporated into development efforts, illustrating how a range of critical engagements with the climate–development interface, while helpfully foregrounding the politics at stake, still operate in ways that validate the centrality of the development model. Section 4 discusses how the work of Sociedade do Sol embeds the logics of post-development carbon – showcasing post-development as a *narrative or intervention*. The chapter ends with a set of conclusions outlining a future research agenda for the coming together of post-development and low carbon urbanism.

2 The post-development critique

In 1995 Arturo Escobar published *Encountering Development*, a critique of development efforts in Asia, Africa and Latin America spanning over half a century. In Escobar's view, these five decades of discourse and practice of development had achieved the very opposite of its aims: underdevelopment, impoverishment, exploitation and oppression. Drawing on Foucault as well as on the works of postcolonial scholars such as Said (1978), Mudimbe (1988) and Bhabha (1990), Escobar challenges notion of development as a historically produced regime of discourse and representation relating forms of knowledge to techniques of power – an apparatus – and operating as a regime of government over the 'Third World'. Development, thus, creates a specific subject people, which ensures Western rule. His analysis builds on other critics of development who had already pointed to the broader geopolitical implications of conceiving the majority of the world's population as 'underdeveloped' (Esteva, 1992), to the technologically led nature of development and its role as an ideological weapon within East–West conflicts (Sachs, 2010 [1992]), and to its false representation as an apolitical project (Ferguson, 1990). This collection of diverse perspectives, referred to as post-development, emerges out of postmodern and post-structuralist critiques and are framed by a decline in the belief in progress and of the role of the nation-state as the entity capable of making it happen (Schuurman, 2000). Early post-development scholars focused on the failings of development, pointing to widespread poverty and growing global inequality despite technological and financial progress in the West (McGregor, 2009). Foregrounding locality and the role of civil society as the sites and agents best placed for the discursive formation of collective futures, Escobar puts forward an interest in identifying 'alternatives to development' (in contrast to development alternatives). Going beyond critique, his work examines how social groups, particularly communities in the global South, "can construct their own social and cultural models in ways not so mediated by a Western episteme and historicity – albeit in an increasingly transnational context" (Escobar, 1995, p. 7).

Post-development perspectives were not received without criticism. Post-development was accused of focusing exclusively on discourse, ignoring the material reality of the very social systems it was examining (Robinson, 2002, p. 1054; see also Nederveen Pieterse, 2000; Rigg, 2003; Corbridge, 1998). Critics defended the development project given its much-needed focus on inequality and the global poor (in contrast to a focus on diversity and plurality, a preferred entry point for post-development scholars). They also suggested that a belief in the retreat of the nation-state (in favour of 'civil society') was naive, while challenging the assumption that 'civil society' would be capable of providing the services provided by the state (Schuurman, 2000). Post-development was blamed for not being able to go beyond critique and failing in the provision of alternative pathways, "offer [ing] no politics besides the self-organising capacity of the poor, which actually lets the development responsibility of states and international institutions off the hook" (Nederveen Pieterse, 2000, p. 187). Its scholars were accused of romanticizing 'soil cultures', drawing on binary anti-modernizing narratives which find little place for the urban in the production of much needed resources (Corbridge, 1998). Even its very intellectual foundations were found at fault. Post-development was accused of misunderstanding power within Foucault, by failing to recognize power as creative and productive, and instead focusing on an oppositional and negative reading of power that spoke of a colonial past rather than a developmental present (Brigg, 2002). The result of such misreading is significant, for it contributes to misplacing agency within the development process: through equating development with the Westernization of the world, post-development "elides the fact that many Third World governments and subjects have actively embraced development" (Brigg, 2002, p. 425; see also Delcore, 2004).

Post-development scholars have responded to (and in some cases incorporated) these critiques, highlighting that their objective is not to get development right, but to construct it as an object "of critique for debate and action" (Escobar, 2012, p. xiv). Since "modernity and capitalism are simultaneously systems of discourse and practice" (Escobar, 2012, p. xiv), post-development's emphasis on discourse, they argue, does not lead to overlooking poverty or material conditions. Finally, taking distance from an oppositional understanding of power, going beyond critique and embracing the agency of the subjects of development, post-development scholars have taken on board – collaboratively with those involved – the task of imagining and bringing about alternatives: in the words of J.K. Gibson-Graham, "the challenge of postdevelopment is not to give up on development, nor to see all development practice – past, present and future, in wealthy and poor countries – as tainted, failed, retrograde … The challenge is to imagine and practice development differently" (Gibson-Graham, 2005, p. 6).

Nearly two decades after its initial formulation, academics and practitioners alike are calling for a renewed engagement with the tenets of post-development, pointing to the possibilities opened by its "post-structural deconstruction of development norms and languages and its concerted attempt to demystify and challenge the taken-for-granted goodness and vagueness" of development (McGregor, 2009, p. 1699). Maintaining an emphasis on foregrounding the political nature of development processes (e.g. Jones, 2004; McGregor, 2005; Kothari, 2005), post-development is empowering silent voices while exploring and enabling unconventional pathways.

It is precisely this emphasis on alternatives that allows post-development today to avoid a debate around what is 'development', focusing instead on what 'post-' could be about (McGregor, 2009). In 2012 Escobar re-edited *Encountering Development*, with a new preface that accounts for a "combined crises of food, energy, climate, and poverty" (Escobar, 2012, p. viii). Escobar points to a plurality of 'Transition Discourses' (TDs) responding to this crises in both the global North and South, seeing them as markers for post-development: avenues where post-development can be imagined and practised. Such TDs, he argues, emerge when the conditions of translation – "the process by which different, often contrasting, cultural-historical experiences are rendered mutually intelligible and commensurable" (Escobar, 2012, p. xx, citing Mezzadra, 2007) – have to be established anew. Offering a broad account of TDs, tracing them to a variety of fields including culture, ecology, religion, spirituality, political economy, alternative science and others, Escobar points to the impossibility of agglomerating these discourses into one single general theory – an epistemology based on a form of pluriversalism rather than universalism.

In this re-articulation of post-development, the 'post-' prefix is interpreted not simply as 'abandoning' or 'moving beyond', but as decentring and displacing. As such, post-development means displacing the discursive and social centrality of development – a space where it no longer operates as the central principle organizing social life. Enriching original post-development debates around resources, ecology and nature, this new post-development draws on the recent experience of Latin American nations such as Bolivia and Ecuador, where, taking into account the cosmologies of the country's indigenous peoples, nature has been recognized as an active subject of community life and its rights enshrined in law (Gudynas, 2009). Echoing Bennett's (2010) call for crossing human–non-human divides and opening possibilities for listening to the vitality of natural matter, post-development affirms a place-based and cultural-ecological framing of human–nature relationships (Escobar, 2008). Alongside calls for the breakdown of the binary nature–culture divide, post-development denaturalizes other deep-seated ontological assumptions that make up a modern universe: the ontological differentiation between individuals and community, the belief in objective knowledge and the recognition of science as the only valid form of knowing. It points to how such a universe finds coherence in socio-material forms such as capitalism and the state. In examining alternative paths, post-development signals the mobilization of strategies to relocalize resources as a way to move from mechanical-industrial paradigms to people- and planet-centred ones (Shiva, 2005, 2008). Critically, it embraces a project of decentring/displacing the capitalist economy while decoupling development and the European capitalist experience of industrial growth. Engaging with Gibson-Graham's feminist critique of political economy (Gibson-Graham, 2006 [1996]), post-development seeks to open up spaces for post-capitalist economies: an economy made of diverse capitalist and non-capitalist practices, and a decentring of capitalism in the definition of the economy. Mobilizing such ontologies means opening possibilities for a post-liberal world: "a space/time under construction when social life is no longer so thoroughly determined by the constructs of economy, individual, instrumental rationality, private property and so forth" (Escobar, 2012, p. xxxi).

So what does post-development mean for low carbon urbanism? This chapter argues that post-development offers conceptual 'yardsticks' to evaluate the extent to which urban low carbon experimentation contributes to envisioning and implementing a pluriverse – characterized by a decentring and displacing of capitalism, by decoupling economy and the European experience of industrial growth, and by decentring liberalism and its associated instrumental rationalities, nature–culture dualisms and bounded construction of the individual. Echoing previous findings of scholars working on low carbon transitions, where a concern with climate change and carbon is always attached to other things (Bulkeley, 2016; Stripple and Bulkeley, 2013), a post-development reading means that 'low carbon' is enacted locally in varied ways, through various name tags and following diverse and sometimes contradictory logics and drivers. But critically, what is at stake in the emerging post-development carbon is a transformation in epistemologies – cultural pluralisms – and ontologies – ways of conceiving nature, individual and society – rather than a change in (development or climate change) paradigms.

3 Development paradigms: thinking about development though climate change

For almost two decades, issues of climate change have been significantly present within development debates. A significant part of this discussion has focused on the linkages between development and climate vulnerability, noting that, as development agencies embrace the logics of adaptation, climate change is rapidly transforming how we think about development (Adger et al., 2003, 2005, 2009). The climate change–development interface has also been considered from the perspective of disaster risk reduction, calling for integrated approaches (Schipper and Pelling, 2006). More recently the links between climate change and development are being examined from a resilience perspective, with research identifying the emergence of a policy discourse that favours the status quo over radical transformation (Brown, 2014; Brown, 2012; Cannon and Müller-Mahn, 2010). Overall, as 'climate development' consolidates as a subfield of research and action, policy-makers are increasingly asked to prioritize climate vulnerability and adaptation policy over issues of GHG reduction and mitigation (Gaillard, 2010; Ayers and Huq, 2009; Ayers and Dodman, 2010; Klein et al., 2007).

Discussions on the interface between climate mitigation and development have been similarly active, albeit to a lesser degree. Taking central stage are debates around who should bear the costs of mitigation and the various ways to finance it (through, for example, the Clean Development Mechanism or other forms of climate finance) (Olsen, 2007; Newell et al., 2009; Disch, 2010; Boyd et al., 2009). Within practitioner and activist communities, the arrival of climate change sparked discussions around 'the right to development' – a debate that, while acknowledging a responsibility for developing countries to reduce GHG emissions, locates the responsibility for funding these efforts within industrialized nations (Baer et al., 2008). Within the dominant understandings of climate mitigation, reducing GHG is largely seen as a co-benefit arising from economic and development interventions, with market interventions recommended as the rightful path to "accelerate economic growth while reducing emissions growth" (Chandler et al., 2002, p. v; see also Davidson et al., 2003).

Development's attempt to integrate mitigation within the development process can be criticized for a limited consideration of its political implications. The reframing of climate change as an opportunity for development often comes via simplistic win–win narratives (Halsnæs et al., 2011; Simon et al., 2012), and through the prioritization of technology transfer (Forsyth, 2007) and market-based mechanisms (Gasper et al., 2013; Boyd and Goodman, 2011). While recognizing a carbon-constrained future and prioritizing the need to curb GHG emissions, dominant debates at the climate–development interface do not challenge development itself, but propose a strategic re-articulation of national-level development priorities around renewable energies or climate friendly industries. What is challenged is the nature of industrial processes, rather than the idea of development itself.

Scholars using political ecology approaches have critically unpacked this climate mitigation–development interface by, for example, pointing to the role that contested politics plays within the governance and financing of clean energy transitions (Baker et al., 2014; Newell et al., 2009). Similarly, political economy perspectives have argued for the inadequacy of technological transfer given the contestable purposes of low carbon development and its neglect of capabilities and contexts (Byrne et al., 2011). Tanner and Allouche (2011) propose a new political economy of climate change and development, asking for greater consideration of ideology, power and resources, while at the same time recognizing the importance of addressing the climate–development interface given its role in achieving poverty reduction as well as other development goals. These critical examinations of forms of mitigation, far from challenging the centrality of the development model, have foregrounded issues of indigenous innovation as well as justice, distribution and labour.

Calls for rethinking development through climate change are more prominent within works on adaptation (see Brooks et al., 2009; Brown, 2011; Pelling, 2010). A growing body of work is challenging the current configuration of the development–climate interface through what is seen as a crisis in capitalism, calling for a reconfiguration of both development and capitalism in order to appropriately respond to the challenge of climate change (Pelling et al., 2012; Storm, 2009; Paterson, 2010). Taking these views one step further, the interface between mitigation, development, justice and equity in climate change has been hailed by scholars familiar with post-development approaches as a possible strategy to transform current economic development models (Harcourt, 2008). Within this intellectual space, both the role of multilateral development agencies such as the World Bank and the mobilization of climate change as a transformative force for development has been called into question. Critiques highlight the way that climate concerns and actions are integrated into neo-liberal logics, neocolonial mythologies and global markets at the service of the global North (Michaelowa and Michaelowa, 2011; Redman, 2008; Lohmann, 2008).

The literature examined above locates the debate primarily within global, international and national scales. Few studies critically unpack the development–climate mitigation interface at subnational domains, although recent analyses are starting to think about what an urban governance of climate change might mean for international development (Castán Broto, 2017). In the case of cities in Asia, Africa and Latin America, analyses of low carbon forms of urbanization have tended to prioritize issues of capabilities, governance structures, measuring instruments and institutional capacity towards low carbon or *sustainable urban development* (Romero-Lankao, 2007; cf. Romero-Lankao and Dodman, 2011; Hoornweg et al., 2011; Revi, 2008; Dhakal, 2010). A limited number of studies have pointed to how urban low carbon initiatives, rather than challenging the ways in which development is imagined, often establish justifications for existing development modes (Bulkeley et al., 2015; Bulkeley and Castán Broto, 2014). A critical strand of studies around low carbon urbanism in cities in the global South has foregrounded the extent to which these interventions are political, by, for example, pointing to the ways in which social and market logics are constantly in negotiation and contestation (Bulkeley et al., 2014) and unpacking the role of participation, resistance and conflict in the making of low carbon urbanism (Aylett, 2010; Castán Broto et al., 2015). While these latter studies challenge the ways in which (urban) development and politics are conceptualized within urban low carbon experiments and interventions, they do not examine in detail how such experiments open up possibilities for difference and envision a configuration of the social that does not revolve around ideas and practices of development. The next section examines how an urban low carbon intervention in the global South reconfigures traditional logics of development and decentres both climate concerns and capitalist economic rationalities in order to mobilize renewable energy towards a variety of social objectives, and in doing so illustrates both the intellectual and practical possibilities of post-development carbon perspectives.

4 DIY SHW systems in São Paulo: illustrating post-development carbon

Since the late 1990s a small non-profit in São Paulo (Brazil) has been working towards the technological development, promotion and implementation of low-cost DIY SHW (do-it-yourself solar hot-water) systems. This energy initiative illustrates post-development carbon logics in an urban setting. In Brazil, throughout the 1990s, an SHW system was seen as an expensive gadget for the rich. In this context, the Sociedade do Sol designed an affordable system that, in their view, is more in tune with the economic and social realities of the country. This low-cost and DIY SHW system is aimed at providing cheap energy for the poor and lower middle classes of the city (Figure 13.1). It can be easily manufactured with PVC (polyvinyl chloride) tubes, plaques and joints available in almost any hardware store. In contrast with commercially available SHW systems, assembling a DIY SHW system does not require specialized knowledge. Sociedade do Sol aims to capture the potential of DIY SHW systems for the promotion of renewable technologies among the public, as well as an opportunity for generating climate awareness and responding to the 'current environmental crisis'. Knowledge about how to manufacture DIY SHW systems is freely distributed over the phone, on

FIGURE 13.1 DIY SHW system installed on the roof of offices of Sociedade do Sol, in
São Paulo, Brazil.
Source: the author.

the Internet and in person by a network of volunteers holding regular workshops,
making presentations in schools and community centres and uploading videos on
YouTube (Figure 13.2).

In Guarulhos, a city of 1.3 million people city located in the São Paulo metropolitan region, the municipal government is supporting the installation of low-cost
DIY SHW systems through the Guarulhos Solar Programme (Law No. 6713, of 1
July 2010). Its main promoter, the Director of the city's Lighting Department,
would like to see 100,000 DIY SHW systems in the city. He cites as the key
reasons for this the support of social agendas through energy technologies, the
reconfiguration of energy subjects in the context of rights and responsibilities, and
the mobilization of local energy initiatives for supporting the local economy. The
local government aims for energy to retain a social function, making energy into an
important element in the "fight against poverty." "It is a solution that comes out of
the poor," says the city's Lighting Director. "Because when Brazilian industry starts
manufacturing solar heaters, what they want is to earn money … the idea of
Sociedade do Sol says the contrary. I am going to put technology at the service of
society, first and foremost, and not at the benefit of private gain" (research interview). According to calculations by the municipality, 100,000 DIY SHW systems
would save 40 megawatts, the same amount of electricity generated by a small to

FIGURE 13.2 Volunteers making a presentation on how to manufacture a DIY SHW system to adult learners at a school in São Paulo.
Source: the author.

mid-size power plant. Capital cost savings would be in the order of 150 million BRL (Brazilian reais), while additional savings could be obtained thanks to the elimination of operational costs and transmission lines. In contrast, they estimate that installing 100,000 DIY SHW systems would require only 60 million BRL, with the majority of this sum coming from micro-investments by the families and businesses installing and using the systems. Each family or business would save in the order of 360 BRL per year on their electricity bill (about 100 GBP (Great Britain pounds)), enabling them to recover the investment in two years. In the view of the municipality, this also represents 36 million BRL that would benefit the economies of families and small businesses, and that could be reinvested in the local economy.

As a socio-technical intervention, the DIY SHW system works under a different set of logics to the way in which traditional energy infrastructures in the city operate. It is underpinned by narratives that seek to bring essential resource flows outside global capitalist circuits and away from profit logics, prioritizing social benefits and community engagement over private gain. Similarly, it is a direct attempt to re-territorialize both natural and financial resources, by fostering local economies around energy flows and empowering dwellers to directly access essential services (such as water heating for showering purposes). DIY SHW systems aim to engage the family in the manufacture and maintenance of the system itself, in effect making the user into a functional piece of the dwelling's energy infrastructure. The founder of Sociedade do Sol highlights a different logic at play when comparing their system with those available in the market, and through that, evidences a different way of conceiving agency within infrastructures as well as a different set of rights around resource flows and infrastructure provision:

> our process is different. Our process is social. It is a process that involves [the user], where the family, the future owner of that house, is the one who has to install it. It is another perspective; not a commercial perspective ... The factories are selling for those who already have houses; not for social projects. The family should participate. It is an obligation. Otherwise, it is not a social project ... Our work is ... to accelerate the pace of spreading the word so that the poorest people start understanding that they have a right to a solar heater ... they need to understand that they have the right to this knowledge.
>
> *(Research interview)*

While logics around low carbon and climate change figure within the overall narrative of Sociedade do Sol, they are not the driving logic; rather they form part of broader narratives around environmental protection, renewable resources, the reconfiguration of production–consumption practices and the reformulation of the economy. In line with accounts around prosumers and the implications of a greater engagement of the user in the infrastructure network (Ritzer et al., 2012; Furlong, 2010), the designers of the DIY SHW system expect users to have, as a result of their active involvement, a greater awareness of resource flows and environmental issues as well as an identification with renewable energy technologies. Infrastructure takes the form of knowledge transmission via social networks, and energy is seen in the context of subject rights (to solar energy) and duties (to transmit solar energy knowledge). The emerging energy arrangement upsets traditional power balances – and its respective financial flows – between central (e.g. national and regional utility companies) and local levels (dwelling and municipality), opening possibilities for new scalar configurations and understandings of the role of the city in relation to the state (Bulkeley, 2005). The subject's right to energy is not exclusively about grid access, a common interpretation that builds on the 'modern infrastructure ideal' (Graham and Marvin, 2001), but about energy affordability via social technologies that lower energy costs. Specifically, in the words of Guarulhos's Lighting Director, rather than being about access to energy, it is about "access to the means of energy production" (research interview).

The experience of the DIY SHW system illustrates a central tenet of a second generation of urban low carbon transition studies – the advancement of novel ways of thinking about the decarbonization–development interface, where low carbon transitions are not simply as a matter of reducing GHG but rather transformative processes that engage with societal forms of development (Luque-Ayala et al., Chapter 2, this volume). It follows that the transformative potential of urban low carbon experiments lays not so much on technological or institutional reconfigurations, but on ways of imagining and implementing collective futures. Inevitably, this transformative potential rests on an acute consideration of the politics that run through the transformation itself. Post-development, as an analytical framing, shifts low carbon urbanism away from the technical or institutional specificity of mitigation and into the new kinds of societal configurations that emerge from certain types of locally grounded experimentation with resource politics. The DIY SHW system, examined through the lens of post-development carbon, decentres both carbon and climate from the experimentation itself. Through this, it opens a space for plural and multiple vernacular interpretations of the *whys* and *hows* of a post-fossil world: from the need to adopt social objectives to a drive to put in place a new relationship with the natural world (e.g. through relying on solar resources); or, from repositioning subjects within the infrastructural configuration of the city to constituting urban infrastructures through collective means and advocating for different ethical and economic relations in the process of accessing and allocating resources.

5 Conclusions

The example of DIY SHW systems in São Paulo, like the iShack in Stellenbosch, represent a type of low carbon urban initiative that opens progressive possibilities for a new set of social and economic agendas associated with climate change mitigation and urban infrastructures in the global South – from poverty alleviation to environmental awareness. It is important to note that not all the 'yardsticks' of post-development are illustrated by these cases (e.g. the ontological reconfiguration of human-nature relations or the possibility of engaging with the vitality of natural matter). However, they illustrate how, through a low carbon initiative, a post-development and post-capitalist order comes into being. In the case of the DIY SHW, this occurs via a system that prioritizes non-market relations, a clear re-territorialization of resource and economic flows (favouring local flows over national and global circulations), and the promotion of novel environmental and 'infrastructural' subjectivities. It is significant that while the low carbon objective was not central to the project, a narrative around an ecological crisis and a purposeful move towards post-fossil fuel arrangements underpins the initiative. Through a post-development carbon lens, the analysis enables the incorporation of what could be seen as 'the vernacular of low carbon': diverse ways by which low carbon logics are interpreted and mobilized in the global South.

Post-development offers an analytical framework that (a) foregrounds how low carbon interventions might provide templates for imagining and enacting a post-fossil fuel world, regardless of how small such interventions are; (b) readily points to the politics embedded within any urban low carbon intervention by calling attention to a discussion around collective futures; and (c) provides markers to examine the extent to which such initiatives transform dominant development modes or operate through the maintenance of the status quo. It provides a conceptual 'yardstick' to evaluate how urban low carbon initiatives contribute to envisioning and implementing a pluriverse capable of decentring capitalism, decoupling economy and growth and transcending liberalism in how we conceive the world. In doing this, it decentres issues of climate mitigation and a reduction in GHG from the very configuration of the low carbon initiative, allowing for a different set of drivers and rationalities to be considered within research on low carbon urbanism. Most importantly, post-development carbon reconciles local priorities in cities in the global South (where a reduction in GHG does not always figure high on the agenda) with action and research on low carbon urbanism.

Post-development carbon opens a new set of research questions for low carbon urbanism. First, the need to deepen our understanding of the political, material and symbolic manifestations of the low carbon–development interface, alongside an inquiry into how such an interface is changing and challenging (or not) traditional understandings of development. This involves developing a critical analysis of how 'development' is understood within urban low carbon initiatives that emerge from within (local, national and international) development efforts, and up to what extent these are operating as tools for domination, for the maintenance of the status quo – or for contestation. It means asking open-ended questions about the shape and form of the vernacular of low carbon in cities in the global South, and examining up to what extent advancing a low carbon agenda might currently be bounded by the scope of possibilities provided by development paradigms. The second set of research questions is methodological, in two ways. From a purely analytical perspective, it is important to ask how we can 'measure' the post-development nature of low carbon urban initiatives, and what does it mean to have conceptual 'yardsticks' for measurement in the context of a theoretical position that embraces the possibility of a pluriverse. From a practical perspective, we need to better understand how and through what means this type of analytical examination of low carbon urbanism can support the advancement of post-development carbon logics on the ground. This potential contribution of both post-development and low carbon urbanism to imagining a post-capitalist and post-liberal society gains a particular salience in the context of rising urban middle classes and an exponential increase in consumption in Africa, Asia and Latin America. Third, in our analysis and practical engagement with low carbon urbanism, we need to find ways to engage with the ontological claims and aspirations of post-development – the breakdown of the binary nature–culture divide, an acknowledgement of the vitality of nature, and the possibility of reconceptualizing the nature of beings and the nature of nature (c.f. Bennett, 2010). This dimension of post-development carbon, absent in the empirical cases examined in this chapter, might open up new understandings of nature in the city as well as of the urban in nature, a much needed endeavour to take forward a low carbon transition.

Acknowledgements

I would like to thank Colin McFarlane, Lucy Baker, Harriet Bulkeley, Vanesa Castán Broto, Ruth Machen and the members of INCUT (Urban Low Carbon

Transitions Network: A Comparative International Network) for insightful comments on earlier drafts of this chapter. Any errors or omissions are my sole responsibility. My deepest thanks also to the many volunteers working with Sociedade do Sol in São Paulo and the members of staff of the municipality of Guarulhos who contributed to the research leading to the chapter.

References

Adger, W.N., Arnell, N.W. and Tompkins, E.L. (2005) Successful adaptation to climate change across scales. *Global Environmental Change*, 15(2), 77–86.

Adger, W.N., Dessai, S., Goulden, M., Hulme, M., Lorenzoni, I., Nelson, D., Naess, L., Wolf, J. and Wreford, A. (2009) Are there social limits to adaptation to climate change? *Climatic change*, 93(3–4), 335–354.

Adger, W.N., Huq, S., Brown, K., Conway, D. and Hulme, M. (2003) Adaptation to climate change in the developing world. *Progress in Development Studies*, 3(3), 179–195.

Ayers, J. and Dodman, D. (2010) Climate change adaptation and development I: the state of the debate. *Progress in Development Studies*, 10(2), 161–168.

Ayers, J. and Huq, S. (2009) Supporting adaptation to climate change: what role for official development assistance? *Development Policy Review*, 27(6), 675–692.

Aylett, A. (2010) Conflict, collaboration and climate change: participatory democracy and urban environmental struggles in Durban, South Africa. *International Journal of Urban and Regional Research*, 34(3), 478–495.

Baer, P., Athanasiou, T., Kartha, S. and Kemp-Benedict, E. (2008) *The Greenhouse Development Rights Framework: The Right to Development in a Climate Constrained World* (2nd ed.). Berlin and Albany, CA: Heinrich Boll Foundation; Christian Aid; EcoEquity; and the Stockholm Environment Institute.

Baker, L., Newell, P. and Phillips, J. (2014) The political economy of energy transitions: the case of South Africa. *New Political Economy*, 19(6), 791–818.

Bennett, J. (2010) *Vibrant Matter: A Political Ecology of Things*. Durham, NC: Duke University Press.

Bhabha, H. (1990) The other question: difference, discrimination, and the discourse of colonialism. In: Ferguson, R., Gever, M., Minh-ha, T.T. and West, C. (eds.) *Out There: Marginalization and Contemporary Cultures*. New York: New Museum of Contemporary Art, pp. 71–89.

Boyd, E. and Goodman, M.K. (2011) The clean development mechanism as ethical development?: reconciling emissions trading and local development. *Journal of International Development*, 23(6), 836–854.

Boyd, E., Grist, N., Juhola, S. and Nelson, V. (2009) Exploring development futures in a changing climate: frontiers for development policy and practice. *Development Policy Review*, 27(6), 659–674.

Brigg, M. (2002) Post-development, Foucault and the colonisation metaphor. *Third World Quarterly*, 23(3), 421–436.

Brooks, N., Grist, N. and Brown, K. (2009) Development futures in the context of climate change: challenging the present and learning from the past. *Development Policy Review*, 27(6), 741–765.

Brown, K. (2011) Sustainable adaptation: an oxymoron? *Climate and Development*, 3(1), 21–31.

Brown, K. (2012) Policy discourses of resilience. In: Pelling, M., Manuel-Navarrete, D. and Redclift, M.R. (eds.) *Climate Change and the Crisis of Capitalism: A Chance to Reclaim Self, Society and Nature*. London: Routledge, pp. 37–50.

Brown, K. (2014) Global environmental change I: A social turn for resilience? *Progress in Human Geography*, 38(1), 107–117.

Bulkeley, H. (2005) Reconfiguring environmental governance: Towards a politics of scales and networks. *Political Geography*, 24(8), 875–902.

Bulkeley, H. (2016) *Accomplishing Climate Governance: New Politics, New Geographies.* Cambridge: Cambridge University Press.

Bulkeley, H. and Castán Broto, V. (2014) Urban experiments and climate change: securing zero carbon development in Bangalore. *Contemporary Social Science*, 9(14), 393–414.

Bulkeley, H., Castán Broto, V. and Edwards, G. (2015) *An Urban Politics of Climate Change.* London: Routledge.

Bulkeley, H., Luque-Ayala, A. and Silver, J. (2014) Housing and the (re) configuration of energy provision in Cape Town and São Paulo: making space for a progressive urban climate politics? *Political Geography*, 40, 25–34.

Byrne, R., Smith, A., Watson, J. and Ockwell, D. (2011) Energy pathways in low-carbon development: from technology transfer to socio-technical transformation. STEPS Working Paper 46. Brighton: STEPS Centre. Available at: http://steps-centre.org/wp-content/uploads/Energy_PathwaysWP1.pdf (accessed 1 August 2017).

Cannon, T. and Müller-Mahn, D. (2010) Vulnerability, resilience and development discourses in context of climate change. *Natural Hazards*, 55(3), 621–635.

Castán Broto, V. (2017) Urban governance and the politics of climate change. *World Development*, 93, 1–15.

Castán Broto, V., Macucule, D., Boyd, E., Ensor, J. and Allenshow, C. (2015) Building collaborative partnerships for climate change action in Maputo, Mozambique. *Environment and Planning A*, 47(3), 571–587.

Chandler, W., Schaeffer, R., Dadi, Z., Shukla, P.R., Davidson, O. and Alpan-Atamer, S. (2002) *Climate Change Mitigation in Developing Countries: Brazil, China, India, Mexico, South Africa, and Turkey.* Arlington, VA: Pew Center on Global Climate Change. Available at: https://www.c2es.org/docUploads/dev_mitigation.pdf (accessed 1 August 2017).

Chatterton, P. (2013) Towards an agenda for post-carbon cities: lessons from Lilac, the UK's first ecological, affordable cohousing community. *International Journal of Urban and Regional Research*, 37(5), 1654–1674.

Corbridge, S. (1998) 'Beneath the pavement only soil': the poverty of post-development. *Journal of Development Studies*, 34(6), 138–148.

Davidson, O., Halsnæs, K., Huq, S., Kok, M., Metz, B., Sokona, Y. and Verhagen, J. (2003) The development and climate nexus: the case of sub-Saharan Africa. *Climate Policy*, 3 (suppl. 1), S97–S113.

Delcore, H.D. (2004) Development and the life story of a Thai farmer leader. *Ethnology*, 43(1), 33–50.

Dhakal, S. (2010) GHG emissions from urbanization and opportunities for urban carbon mitigation. *Current Opinion in Environmental Sustainability*, 2(4), 277–283.

Disch, D. (2010) A comparative analysis of the 'development dividend' of Clean Development Mechanism projects in six host countries. *Climate and Development*, 2(1), 50–64.

Escobar, A. (1995) *Encountering development: the making and unmaking of the third world.* Princeton, NJ: Princeton University Press.

Escobar, A. (2008) *Territories of Difference: Place, Movements, Life, Redes.* Durham, NC: Duke University Press.

Escobar, A. (2012) *Encountering Development* (Re-issue with a new preface by the author). Princeton, NJ: Princeton University Press.

Esteva, G. (1992) Development. In: Sachs, W. (ed.) *The Development Dictionary.* London: Zed Books, pp. 6–25.

Ferguson, J. (1990) *The Anti-politics Machine: 'Development', Depoliticization, and Bureaucratic Power in Lesotho*. Cambridge: Cambridge University Press.

Forsyth, T. (2007) Promoting the 'development dividend' of climate technology transfer: can cross-sector partnerships help? *World Development*, 35(10), 1684–1698.

Furlong, K. (2010) Small technologies, big change: rethinking infrastructure through STS and geography. *Progress in Human Geography*, 35(4), 460–482.

Gaillard, J.-C. (2010) Vulnerability, capacity and resilience: perspectives for climate and development policy. *Journal of International Development*, 22(2), 218–232.

Gasper, D., Portocarrero, A. and Clair, A. (2013) The framing of climate change and development: a comparative analysis of the Human Development Report 2007/8 and the *World Development Report 2010*. *Global Environmental Change*, 23(1), 28–39.

Gibson-Graham, J.K. (2005) Surplus possibilities: postdevelopment and community economies. *Singapore Journal of Tropical Geography*, 26(1), 4–26.

Gibson-Graham, J.K. (2006 [1996]) *The End of Capitalism (as We Knew It): A Feminist Critique of Political Economy*. Minneapolis: University of Minnesota Press.

Graham, S. and Marvin, S. (2001) *Splintering Urbanism: Networked Infrastructures, Technological Mobilities and the Urban Condition*. London: Routledge.

Gudynas, E. (2009) *El mandato ecológico: derechos de la naturaleza y políticas ambientales en la nueva Constitución*. Quito: Abya Yala.

Halsnæs, K., Markandya, A. and Shukla, P. (2011) Introduction: sustainable development, energy, and climate change. *World Development*, 39(6), 983–986.

Harcourt, W. (2008) Walk the talk – putting climate justice into action (editorial). *Development*, 51(3), 307–309.

Hollands, R. (2015) Critical interventions into the corporate smart city. *Cambridge Journal of Regions, Economy and Society*, 8(1), 61–77.

Hoornweg, D., Sugar, L. and Gomez, C. (2011) Cities and greenhouse gas emissions: moving forward. *Environment and Urbanization*, 23(1), 207–227.

Jones, P. (2004) When 'development'devastates: donor discourses, access to HIV/AIDS treatment in Africa and rethinking the landscape of development. *Third World Quarterly*, 25(2), 385–404.

Klein, R.J., Eriksen, S.E., Næss, L.O., Hammill, A., Tanner, T., Robledo, C. and O'Brien, K. (2007) Portfolio screening to support the mainstreaming of adaptation to climate change into development assistance. *Climatic Change*, 84(1), 23–44.

Kothari, U. (2005) Authority and expertise: the professionalisation of international development and the ordering of dissent. *Antipode*, 37(3), 425–446.

Lohmann, L. (2008) Carbon trading, climate justice and the production of ignorance: ten examples. *Development*, 51(3), 359–365.

McGregor, A. (2005) Geopolitics and human rights: unpacking Australia's Burma. *Singapore Journal of Tropical Geography*, 26(2), 191–211.

McGregor, A. (2009) New possibilities? shifts in post-development theory and practice. *Geography Compass*, 3(5), 1688–1702.

Mezzadra, S. (2007) Living in transition: toward a heterolingual theory of the multitude. EIPCP multilingual webjournal. Available at: http://eipcp.net/transversal/1107/mezzadra/en (accessed 5 December 2017).

Michaelowa, A. and Michaelowa, K. (2011) Climate business for poverty reduction? The role of the World Bank. *Review of International Organizations*, 6(3–4), 259–286.

Mitchell, T. (2011) *Carbon Democracy: Political Power in the Age of Oil*. New York: Verso Books.

Mudimbe, V.Y. (1988) *The Invention of Africa: Prognosis, Philosophy and the Order of Knowlegde*. Indianapolis: Indiana University Press.

Nederveen Pieterse, J. (2000) After post-development. *Third World Quarterly*, 21(2), 175–191.

Newell, P., Jenner, N. and Baker, L. (2009) Governing clean development: a framework for analysis. *Development Policy Review*, 27(6), 717–739.

Olsen, K.H. (2007) The clean development mechanism's contribution to sustainable development: a review of the literature. *Climatic Change*, 84(1), 59–73.

Paterson, M. (2010) Legitimation and accumulation in climate change governance. *New Political Economy*, 15(3), 345–368.

Pelling, M. (2010) *Adaptation to Climate Change: From Resilience to Transformation*. London: Routledge.

Pelling, M., Manuel-Navarrete, D. and Redclift, M. (2012) *Climate Change and the Crisis of Capitalism: A Chance to Reclaim, Self, Society and Nature*. London: Routledge.

Redman, J. (2008) *World Bank: Climate Profiteer*. Washington, DC: Sustainable Energy and Economy Network; Institute of Policy Studies.

Revi, A. (2008) Climate change risk: an adaptation and mitigation agenda for Indian cities. *Environment and Urbanization*, 20(1), 207–229.

Rigg, J. (2003) *Southeast Asia: The Human Landscape of Modernization and Development*. London: Routledge.

Ritzer, G., Dean, P. and Jurgenson, N. (2012) The coming of age of the prosumer. *American Behavioral Scientist*, 56(4), 379–398.

Robinson, W.I. (2002) Remapping development in light of globalisation: from a territorial to a social cartography. *Third World Quarterly*, 23(6), 1047–1071.

Romero-Lankao, P. (2007) Are we missing the point?: particularities of urbanization, sustainability and carbon emissions in Latin American cities. *Environment and Urbanization*, 19(1), 159–175.

Romero-Lankao, P. and Dodman, D. (2011) Cities in transition: transforming urban centers from hotbeds of GHG emissions and vulnerability to seedbeds of sustainability and resilience: introduction and editorial overview. *Current Opinion in Environmental Sustainability*, 3(3), 113–120.

Sachs, W. (2010 [1992]) *The Development Dictionary: A Guide to Knowledge as Power*. London: Zed Books.

Said, E. (1978) *Orientalism*. New York: Random House.

Schipper, L. and Pelling, M. (2006) Disaster risk, climate change and international development: scope for, and challenges to, integration. *Disasters*, 30(1), 19–38.

Schuurman, F.J. (2000) Paradigms lost, paradigms regained? Development studies in the twenty-first century. *Third World Quarterly*, 21(1), 7–20.

Shiva, V. (2005) *Earth Democracy: Justice, Sustainability and Peace*. London: Zed Books.

Shiva, V. (2008) *Soil Not Oil: Climate Change, Peak Oil and Food Insecurity*. London: Zed Books.

Simon, G.L., Bumpus, A.G. and Mann, P. (2012) Win–win scenarios at the climate–development interface: challenges and opportunities for stove replacement programs through carbon finance. *Global Environmental Change*, 22(1), 275–287.

Storm, S. (2009) Capitalism and climate change: can the invisible hand adjust the natural thermostat? *Development and Change*, 40(6), 1011–1038.

Stripple, J. and Bulkeley, H. (2013) Introduction: on governmentality and climate change. In: Stripple, J. and Bulkeley, H. (eds.) *Governing the Climate: New Approaches to Rationality, Power and Politics*. Cambridge: Cambridge University Press, pp. 1–26.

Sustainability Institute (n.d.) iShack Project. Available at: http://www.sustainabilityinstitute.net/programmes/ishack (accessed 1 August 2017).

Swilling, M. (2014) Rethinking the science–policy interface in South Africa: experiments in knowledge co-production. *South African Journal of Science*, 110(5–6), 1–7.

Tanner, T. and Allouche, J. (2011) Towards a new political economy of climate change and development. *IDS Bulletin*, 42(3), 1–14.

14

CONCLUSION

Simon Marvin, Andrés Luque-Ayala and Harriet Bulkeley

The conclusions are organized thematically following the three-part structure of the book, first examining the *technologies, materialities and infrastructures* of the urban low carbon transition, second the various ways in which *intermediation and governance* arrangements are at play and thirdly the ways in which *communities and subjectivities* become central in the making of low carbon urbanism. In looking at these three themes, we summarize the key arguments of the various chapters included in the book, identify the key research findings and then highlight the implications for further research on urban low carbon transitions. At the end of each section we synthesize the key contribution of the part to the overall aim of the book.

Part I: Technologies, materialities, infrastructures

The first part of the book examined the objects and flows of the material world involved in the production or withdrawal of carbon, and the set of mechanisms and techniques that operate as material, framing and discursive devices capable of influencing both agents and objects. The four chapters of Part I examined a variety of techno-material entry points to understanding both low carbon and energy transitions in Paris, Berlin and Hong Kong, Manchester and a selection of North American cities.

In Chapter 3, Jonathan Rutherford examined, through a case study of Paris, the diverse array of material sites, artefacts and contexts in which low carbon transitions become contested and politicized. The key research finding is that there is a critical gap between the formal low carbon strategies and what is achieved in practice. Yet this is more than the familiar 'implementation' gap. Instead it even concerns the gap between what formal strategies claim to measure as objectives and their disconnections from a wide range of hidden and unrecorded events. Central to this is understanding the ways in which the non-human agency of materials becomes embedded in political processes. This is illustrated through three tensions. First, as more information is provided about the functionality or performance of materials, more claims become contested; second, materials often become subject to claims that exceed their capacity to meet them; and third, materials can be recalcitrant and behave in unruly ways. The implication for further work on urban low carbon transitions is the need to acknowledge that the material world often escapes full control and in complex socio-technical systems becomes an actor – a recombinant agency.

In Chapter 4 Timothy Moss and Maria Francesch-Huidobro examined the fascinating history of the electric city in Berlin and Hong Kong in response to geopolitical isolation and the search for, and limits of, energy autonomy. Critically these show how imposed autonomy shaped the boundaries of the system, despite a remarkable continuity in the socio-technical organization of the network. The key finding in the post-unification context is that the legacy of autonomy is not lost but is developed in subtle ways in each of the different contexts. In each context, the current debates about low carbon transition are powerfully framed through the specific history of the energy system but not determined by it. Further research on urban low carbon transitions could be more closely informed by the macro- and micro-political histories of infrastructural configurations, and explore the ways in which these trajectories may materially and socially shape the pathways of low carbon development.

In Chapter 5 the analysis provided by Mike Hodson, Simon Marvin and Andy McMeekin examined the symbolic rise and relative decline of low carbon transition in the city region of Greater Manchester, and the critical role of the state in conditioning these local responses. The key research finding points to the way in which urban responses are powerfully shaped by wider politically produced economic priorities. Key to this was the positioning of the city region as a national test bed for a largely privately led low carbon transition, shadowed by the relative decline of this agenda as state priorities shifted. The key implication for further research is to examine how urban low carbon transitions are themselves conditioned – through constraint and enablement – in response to state priorities and the need to explore under what circumstances alternative pathways can be developed that may be in tension with state priorities.

Laura Tozer, in Chapter 6, examined a range of local authorities in North America to illustrate variation and difference within discourses on 'carbon neutrality'. The key finding is that there are significant differences in thinking about low carbon – and these are powerfully shaped by the particular social and material context within which the response is developed. Yet the material changes envisioned through the transition are remarkably similar. The infrastructural shape and form of carbon neutral futures usually encompasses highly energy-efficient buildings, solar energy and district heating and energy demand management. Although there may be different ways to be low carbon, there are particular material and social configurations that are favoured. These configurations are not path dependent but do have agency and are path shaping. The key issue for further research on urban transitions is to develop new ways of imagining low carbon futures that may be able to exceed the limits of these configurations.

Part I then reveals three critical issues that need to be considered more widely in the field of urban transitions research. The first is the role of social and political issues – including governance framework and historical context – in shaping the urban priorities and the construction of low carbon transitions pathways. The contributions graphically illustrate the ways in which these contexts condition, but do not determine, the priorities and organization of low carbon transitions. Undoubtedly context is a powerful factor in shaping what is possible to consider as the social organization and priorities of urban strategies. The second issue is understanding the recombinant agency of specific materials and broader technological configurations in shaping the constitution of urban transitions. The contributions raise two dimensions that need to be the focus of future work. On the one hand, despite the diversity of different visions of low carbon transitions there is often a remarkable consistency in the technological and social configurations that are selected for realizing the transition. This paradox needs further analysis in order to work through why, despite the diversity of visions and contexts, a particular package of technological options is selected and what options are excluded. More widely the unruly agency of materials requires further analytical and empirical work across different contexts to uncover the work that non-human materials do in constraining and opening the potential for realizing urban transitions. The third contribution is the need to look outside and beyond conventional understandings of urban low carbon pathways in order to enlarge the purpose, inclusivity and socio-technical options of currently unimagined urban transitions pathways. To unlock the constrained and often path-dependent logic of existing pathways there is need to look towards science fiction and even film and theatre, for ways of envisioning and imaging an enlarged range of urban futures.

Part II: Intermediation and governance

Part II looked at a broad range of activities and structures that make up climate responses at the local level, with a specific focus on municipal governance arrangements and intermediation practices. The four chapters that make up this part advance novel conceptual and empirical ways of examining the role of local governments and a variety of forms of local intermediation in the making of urban low carbon transitions. They are based on in-depth empirical cases from Sweden, India, the state of Victoria in Australia, and African cities.

In Chapter 7, Susie Moloney and Ralph Horne examined the role of intermediaries in enabling low carbon experiments in Australia – a context where municipal authorities lack the capacity and resources to enact a low carbon transition. The key finding is that these regional intermediaries undertake three roles: they develop new forms of experimentation, undertake capacity-building with multiple stakeholders and then seek to upscale experiments through other strategies and networks. But there are limitations on these responses. They tend to be rather exclusive, with a focus on government and business rather than civil society; they are primarily focused on technological solutions, with less attention on social change; and the wider structure of carbon-intensive industries and lifestyles is not addressed. A key issue for further research on urban transformations is whether different modes of intermediation can be less focused on business as usual and have a more transformative focus.

In Chapter 8, Mikael Granberg examined how a municipality developed an intermediary role in the financing of a market-based low carbon urbanism in Sweden. This was designed to finance low carbon businesses through green bonds – including low carbon and clean technologies, energy efficiency and renewables – through partnerships and investment in private businesses. The objective was to accelerate emissions reduction, create investor profit and generate economic surplus for the municipality. Yet the chapter shows that in the short term these differing and complex objectives were extremely difficult to meet and that under current conditions the municipality was having to subsidize businesses. The key issue this raises is the challenges involved in developing intermediaries that attempt to combine both market and collective values that are extremely difficult to hold together in a single entity. More widely the chapter raises further research questions about wider issues of accountability and control of intermediaries that operate in both a market and municipal context.

The final two chapters of Part II look specifically at intermediation and governance capacity for low carbon in the context of cities in the global South. In Chapter 9 Jonathan Silver and Simon Marvin point to the contested politics inherent in the governing of urban energy networks in sub-Saharan Africa. The key finding of the chapter is that there are significant limits in urban governance, capacity and knowledge needed to understand and reshape urban energy networks. An understanding of the urban energy system requires recognition of the diverse material landscapes of energy provision, the low level of access to formal energy networks, the critical role of informality and the potential role of the agency of civil society in shaping energy transitions. The chapter highlights the importance of understanding both context and the political tensions between different social, ecological and economic outcomes of low carbon pathways. In Chapter 10, Neha Sami examined the complex and at times muddled environmental governance system present at regional and local levels in India. The key finding is that environmental governance is framed through the priorities of development and economic growth. Urban capacity for local climate action is notably absent in this context. This provides a marked contrast to the narrative that climate action should be located at the urban level.

Part II makes two critical contributions to future research priorities on low carbon transitions. The first concerns the role of intermediaries and the tendency for these to make up for the deficits of the existing fragmented landscape of local government and the constraints on enlarging the functional role of local authorities. Yet both chapters found rather limited scope of the low carbon transition envisioned by these intermediaries with their rather exclusive focus, giving primacy to techno-economic solutions and working with the existing societal context and constraints. The critical question this raises is whether intermediaries are able to effect more transformative change by developing more inclusive approaches, contributing to change social practices and cultural change and questioning the dominant logic of carbon-intensive industries. The second key contribution is questioning the claims made about the critical role of local government and urban authorities in responding to climate change. Both contributions on African and India graphically show that local authorities have tightly restricted roles and responsibilities to engage in the development of urban energy and low carbon transitions. Centrally these point to the importance of looking beyond local government to other organizational contexts – including the home, civil society and international agencies – that can develop capacity to shape transitions in an urban context.

Part III: Communities and subjectivities

The third and final section of the book focuses on the role of individuals and communities within the low carbon urban transition. The chapters included here look at the increasing government focus on conducting carbon conducts, as well as the collective processes of developing and enacting shared low carbon identities. It is composed of three chapters looking at the experiences of various Australian cities, Ottawa and São Paulo.

In Chapter 11, Robyn Dowling, Pauline McGuirk and Harriet Bulkeley report on the organization of urban low carbon behaviour change initiatives in Australian cities. The chapter outlines an ecology of governing carbon conduct through three modes – rational subject, self-disciplining and material engagement through energy objects. The chapter shows how these different approaches are assembled in novel and varied configurations. The key contribution for further research is twofold. First, the need to move beyond an individual focus in behaviour change programmes and to also consider the collective self through moral motivations in place-based communities. Second, in echoing the findings in Part I, the need to focus on the materials associated with low carbon urbanism as a way of mobilizing new ways of moving beyond individual behaviour change.

In Chapter 12 Matthew Paterson and Merissa Mueller reveal how political conflicts around the future of the city are expressed through fundamental ambivalences in a normal city – one that is not conventionally thought of as a low carbon exemplar: Ottawa. Despite local support for 'liveable city' approaches and the need to address climate change, opposition to intensification and densification of urban development stands in the way of a set of interventions that might favour low carbon urbanism. The distinctiveness of the chapter is that it does not deal directly with the explicit politics of low carbon but instead shows how other processes create contradictory logics that close down the space for low carbon experimentation. The key implication for further research is to look outside the exemplary cities that constitute low carbon responses, to see how conventional urban politics in the ordinary city can close down and potentially opened-up spaces for debating a low carbon transition.

In the final chapter of the book, Chapter 13, Andrés Luque-Ayala reflects on the possible contribution of post-development perspectives to contemporary debates on low carbon urbanism. The chapter shows how a post-development logic can construct a vernacular of low carbon that builds into other priorities by: imagining new post-fossil fuel futures; focusing on collective futures with a wider range of social and cultural objectives; and moving beyond the status quo to consider other possibilities beyond capitalism. Looking forward to future research, the priority is to consider whether low carbon transitions are part of a logic of domination, imposition and maintenance of the status quo, and whether this can be contested and new possibilities developed. Can low carbon be used as a way of exploring new imaginaries of a post-capitalist and post-fossil fuel society.

Part III of the book then tells us three things about the future development of urban low carbon transitions research priorities. First, the potential of a move away from an individual focus (e.g. as part of behavioural change programmes) towards a consideration of the ways in which both people and their relationships with materials operate in wider societal contexts. There is more progressive potential in considering the collective context for individual change and the ways in which a focus on energy and materiality can develop novel and innovative hybrid formations of behavioural change no longer solely focused on individuals. Second, the need to create a shift away from the focus on exemplary urban sites to consider the ways in which normal or ordinary cities may shape low carbon responses. Key to this may be looking outside a conventional carbon focus to the ways in which existing urban politics may delimit and enlarge in different circumstances the potential for progressive carbon politics. Third, the need to consider an urban carbon politics that may move beyond a focus on business as usual, working across existing social and economic constraints to consider what a post-capitalist low carbon society might look like. As pointed out in Part I and Part II, the task here is to examine how the concept of low carbon can be reimagined in progressive ways that can open new possibilities and configurations.

The book has set out different pathways for the city to become low carbon, decentring the central tenet of lowering GHG emissions and foregrounding ways of thinking about development and collective futures. The low carbon transition is not a singular problem, but rather a problem full of multiplicities, meanings and potential responses. As we have seen in over two decades of action and research at the interface between cities and climate change, it is a problem that is continually in transformation, from one of environmental protection and decarbonization to one of green growth, economic resilience, development logics and more.

Yet, in reflecting about the next two decades, there is more to be done in thinking about the next phases of work on urban low carbon transition. Looking over the contributions and themes developed in this book, we argue that there are three key challenging priorities. First, the need to consider urban sites in both the global North and South that are not considered climate exemplars. Here we might seek to reconsider the role of normal and ordinary cities and contexts that lack the capacity and resources to become active in low carbon transitions. In these cities, we may seek to look beyond the municipal level to consider a wider range of contexts and sites where a politics of low carbon transition could be developed. Contributors in this book have shown how it is necessary to challenge the conceived and often repeated wisdom that urban responses can act on climate change – when in fact the picture is much more varied and differentiated. Second, there is the issue of lock-in and obduracy. While on the one hand the book has illustrated the diversity of urban low carbon visions, there is a tendency for responses to focus on a particular suite of technological (rather than social) solutions, to give primacy to business and elites (rather than civil society), and to work within existing social and economic constraints (rather than question or exceed them). This pattern is often repeated throughout the chapters. The critical question is, then, can this constrained response be opened and problematized with the potential for new solutions? The third issue the book has consistently raised is the need to develop contexts, techniques and processes that can help to enlarge and open how we think about development, expanding the range of new imaginaries of what a low carbon future might look like. Our authors have consistently grappled with the need to move beyond the status quo and business as usual, and all have argued in favour of a search for new visions and ideas that can populate what a post-capitalist, -development and -carbon society might look like. Much is at stake here. This would mean the research community engaging in new relationships and collaborations with civil society, grass-roots organizations, collectives, artists, writers, film-makers and a range of other creative spirits in the development of new methods and processes for envisioning and experimenting with different urban futures.

INDEX